Natalie Beckerman knows what it truly means to provide meaningful service. In *When Did You Stop Caring?*, she describes the true spirit of hospitality and brings it together with the power of process and measurement. This book is a service design guide for building loyalty, empowering your team, and creating long-lasting customer memories. Every sales, service, loyalty, and customer experience leader should read this.

—**Becky Ploeger**, global head of reservations, customer care, and sales support at Hilton

Society is rapidly sacrificing our humanity on the altar of efficiency. Natalie's book and challenge is the timeout we desperately need to get back to treating people as human beings, not as human doings.

—**Davin Salvagno**, #1 bestselling author of *Thieves of Purpose*

In an era of AI and automation, Natalie Beckerman reminds us that humanity remains the ultimate differentiator. *When Did You Stop Caring?* challenges leaders to use technology to elevate the human connection, not to replace it. This is a must-read for modern CX leaders.

—**Kali Sarim**, cofounder and CEO of InfiniteCX

Natalie Beckerman brings a rare combination of heart and operational rigor to the customer experience conversation. *When Did You Stop Caring?* is a bold challenge to leaders to rethink how they serve, lead, and grow. This book will shape the future of customer experience.

—**Mario Matulich,** president, customer management practice

As a CEO, I've experienced how the pressure of pursuing tech-driven efficiencies unintentionally strips away humanity. Natalie Beckerman has articulated what many leaders struggle to confront: Trust erodes when care disappears from our processes, and performance often follows. *When Did You Stop Caring?* is a welcome challenge to lead with intention in building organizations that customers trust and employees are proud of.

—**Michael Joubert**, founder of The Billybo Group, active board member of several organizations, former chairman of Saatchi & Saatchi, and EVP of marketing for Levi Strauss & Co

PRAISE FOR *WHEN DID YOU STOP CARING?*

When Did You Stop Caring? delivers a hard lesson about the importance of humanity. Natalie demonstrates how to get real results from listening, bold thinking, deep caring, and enthusiastic delivery. Natalie also provides a solid argument for achieving success with the right mix of culture, AI, and successful BPO relationships. You must read this book.

—**Chris Crowley**, president and CEO of iQor

Caring is your competitive advantage. This book shows leaders how to build a culture where care drives action, loyalty, and results. When employees care, customers feel it, and they come back. For any leader looking to create loyalty among customers and engagement among employees, this book should be your first stop.

—**Shep Hyken**, customer service and CX expert, keynote speaker, and *New York Times* bestselling author

When Did You Stop Caring? is a sharp, necessary challenge to leaders who've mistaken efficiency for excellence. In an era when AI promises to automate everything, the leaders who will win are the ones who don't forget to lead with the customer experience and work backward from there. Because the algorithm doesn't care. You still have to.

—**John Rossman**, author of the *Amazon Way* and *Big Bet Leadership*

Nobody has the right to be in business. You must earn and re-earn it every day—in every encounter, with each of your customers. If you believe this business principle, you must read, absorb, and implement all that Natalie Beckerman teaches in *When Did You Stop Caring?*

—**Scott Miller**, 8X bestselling author, former host of the world's largest weekly leadership podcast, former CMO and EVP of FranklinCovey keynoter, principal at the Gray + Miller Agency

When did we decide that efficiency mattered more than dignity? Natalie Beckerman's compelling book is a wake-up call to leaders who've lost themselves in spreadsheet optimization and forgotten the human! Caring isn't soft—it's strategy. Culture is the operating system and service the brand. Technology's role is to amplify humanity not replace it.

Natalie shows that if you build for trust and empower your people, then loyalty, growth and relevance will follow.

—**Ronan Dunne**, chairman and strategic advisor, former CEO, Verizon Wireless and O2 Telefonica

Natalie Beckerman presents an impassioned argument that values and performance aren't mutually exclusive. They're linked as one. *When Did You Stop Caring?* illustrates how to build lasting loyalty through principled leadership that has significant business implications for purpose-driven leaders.

—**Michael Kieschnick**, former CEO of CREDO Mobile and president and founder at Green Advocacy Project

It's no small act of courage to write a book about caring in a climate that often celebrates urgency over empathy and profit over principle. Yet, that is exactly what Natalie Beckerman has done! She challenges leaders to rethink what strength really looks like and to recognize that compassion and performance are not opposites. They are partners.

—**Richard Eychner**, chief culture officer at iQor

When Did You Stop Caring? provides an easy roadmap to develop empathetic leaders by modeling the right behaviors yourself and create a great culture by teaching your people about the impact of their daily behaviors instead of simply teaching them how to perform. Natalie Beckerman provides business leaders with the exact tools needed to deliver real ROI from caring about your customers and your people serving your customers.

—**Dan McCann**, CEO and chief learning officer at SymTrain

Great service does not have to be costly; being caring has ROI, and its impact can be measurable, scalable, and even profitable. Natalie Beckerman shows us how AI and technology can amplify and enhance human connection, not replace it. This book provides the missing link between data, discipline, and the human connection we all need in the age of AI.

—**Dave Rennyson**, president and CEO of SuccessKPI, Inc

As a leader of a BPO, I know that it is people, not just tech & efficiency, that drive loyalty. Natalie emphasizes that AI should complement and build upon the "human" element, rather than replace it. She provides a practical and inspirational roadmap to building service-based businesses that deliver high performance while being rooted in a caring culture.

—**Robin Hoekstra**, CEO of Outworx

"You can't go back and change the beginning, but you can start where you are and change the ending."

—C.S. Lewis

MAISON VERO

Published by Maison Vero
1619 Providence Rd., Ste 220-254
Marvin, NC 28173

Inquiries may be directed to: Maison Vero, 1619 Providence Rd., Ste 220-254 Marvin, NC 28173, or info@graymilleragency.com.

For information about special discounts for bulk purchases, please call 1-949-333-4872 or email info@graymilleragency.com.

Maison Vero is a partner brand of The Gray + Miller Agency, a speaking, literary, and talent consortium.

For more information on the talent represented by The Gray + Miller Agency, or to bring any of our thought leaders to your organization or live event, please visit our website at graymilleragency.com

Cover Design: Mike Elwell
Book Design: Mel Wise

Manufactured in the United States of America

Paperback: 978-1-969508-45-5
E-book: 978-1-969508-46-2
Hardcover: 978-1-969508-44-8

TABLE OF CONTENTS

PART III—WHERE INTENTION MEETS EXECUTION

PART IV—THE BUSINESS CASE FOR CARING

PART V—BUILDING A COMPANY THAT LASTS

PART VI—REIGNITE CARING AND FAN THE FLAME

FOREWORD

GARRY RIDGE

Chairman Emeritus, WD-40 Company, The Culture Coach,
and Bestselling author of *Any Dumb Ass Can Do It*

I've spent most of my life inside organizations. Some worked beautifully. Some didn't. And here's what I've learned the hard way.

When things go wrong in a business, it's rarely because people don't care.

It's because leaders stopped caring first.

Not intentionally. Not maliciously. Just gradually. Quietly. One policy at a time. One efficiency decision at a time. One "we'll fix that later" conversation at a time.

And before you know it, caring becomes optional.
That's why this book matters.

When Did You Stop Caring? doesn't sugar-coat anything. It doesn't hide behind buzzwords. And it doesn't let leadership off the hook. It holds up a mirror and asks the question every leader needs the courage to answer honestly: When did you stop caring about customers as human beings?

When did policies become more important than dignity? When did efficiency replace empathy?

I've seen what happens when leaders choose care. During my leadership at WD-40 Company, we learned something simple but powerful: You don't build great companies by squeezing people. You build them by trusting people, helping them learn, and creating environments where they feel safe, valued, and respected. When people feel that way at work, they show up that way for customers.

Caring is not soft. It is not fluffy. And it is definitely not naïve. Caring is a leadership discipline.

This book makes that point clearly and relentlessly. It shows how culture becomes your operating system. How leaders set the emotional thermostat. How policies can either help people get an A or mark their paper. And how technology, when misused, can scale indifference faster than any human ever could.

What I appreciate most about this book is that it doesn't just diagnose the problem. It gets practical. It shows how caring shows up in hiring, training, policies, service recovery, and everyday decisions. Not in theory. In real life. With real consequences.

And here's the uncomfortable truth every leader needs to hear. If your customers feel unheard, unseen, or disrespected, that didn't start in the contact center. If your employees feel powerless, burnt out, or afraid to do the right thing, that didn't start with them either.

That started with leadership.

This book is not about being nice. It's about being effective. It's about understanding that trust, loyalty, and long-term success are built one human interaction at a time. And that leaders, whether they like it or not, are always teaching people what matters most by what they reward, tolerate, and model.

If you're looking for a quick fix or a shiny framework, this book may make you uncomfortable.

Good. Growth usually does.

But if you believe, as I do, that leadership is about learning and teaching, that culture is shaped by behavior not posters, and that businesses exist to serve people, not the other way around, then this book will resonate deeply.

So read it with an open mind. Read it with humility. And most importantly, read it willing to ask yourself the hardest question of all: Am I being the leader I want to be right now?

One thing that Natalie so eloquently articulates is that AI can never replace your humanity.

Because caring doesn't start with customers. It starts with us.
Cheers, **Garry Ridge**

INTRODUCTION:

THE MOMENT WE FORGOT HOW TO CARE

I have had a deep appreciation for—and even a bit of an obsession with—exceptional customer service since I was young. I was only six years old when I first realized how important it is to treat customers with respect, trust, and care.

Even then, I sensed something many leaders still struggle to grasp for any brand or business to thrive long-term: It must make its customers feel genuinely valued, respected, and trusted. This is not a soft philosophy. It is hard business.

The memory where this realization happened is still vivid. I had been on a routine grocery shopping trip with my mother, which, as a kid, was as painful as a visit to the dentist. I hated every second of grocery shopping. To survive these outings and make this torture bearable, I created my own version of entertainment. I'd pick up a handful of peanuts from the large container barrels, and whenever my mom would stop to check an item, I would discreetly drop a shelled peanut underneath her feet.

As she stepped back, she would smash the peanut, and it would explode into a loud crunch. I would do this again and again. Behind us, a trail of crushed peanut shells and minor destruction marked our slow progress through the store. With each step of hers, my excitement and amusement increased. "We just need a few more items," my mom would say, and I would intensify my clandestine operation, perfecting my precision peanut-squashing attempts to keep her from dawdling. It did leave a significant amount of peanut shell carnage trails throughout the store.

As my mom placed each item on the conveyor belt with quiet confidence, she set each coupon alongside it, lining them up with their matching product like a teacher grading their own perfect work. Each coupon was valid. Each detail matched.

As we finally entered the checkout area, the fluorescent lights hummed above, the conveyor belt rolled with a rhythmic motion... but there was a heaviness in the air. Something felt off.

Now, my mom was an extraordinary Sunday newspaper coupon-clipping expert. It was a tradition in our house—and how my mom made the most of every dollar without losing her pride. Over the years, I watched as she sat at the kitchen table for hours, scissors in hand, cutting each rectangle with care, placing them in neat stacks, and organizing them by brand, by type, and by expiration date and fine print. Coupons were more than a means of saving money. They represented discipline, care, and responsibility.

The first beep from the scanner that day was common, but the second was delayed. Then the third was sharp. The cashier's eyes narrowed as she examined the first coupon. She looked at it for a moment longer than she needed to, flipped it over, inspecting it as if it were fake. Her demeanor was cold. "Are you sure this applies to this item?" she asked suspiciously.

My mom softly leaned forward "Yes," she replied calmly, "it specifically says that in the fine print."

The cashier scanned it, but with a sigh of exasperation. The sound of the beep that followed was different—it felt accusatory. Each item that crossed that red laser increased the tension. Each coupon was subject to scrutiny. She evaluated each one individually, her voice growing tighter and her body position becoming stiffer. The exchange that should have taken only minutes to complete quickly turned into something much more awkward. The line behind us was getting longer. Carts were shifting. Someone coughed impatiently.

This occurred long before self-service checkout counters and kiosks. To provide service was to interact with humans. To serve customers meant to look at them, to show them respect. At least that is what was supposed to happen. There was supposed to be some level of dignity involved in the transaction.

My mom was still composed and cooperative—more than she should have been. She followed every direction. She answered every question. She was courteous, polite, and honest to a fault. Yet none of these characteristics caused the cashier to soften her tone. If anything, she became even more aggressive. What began as a cautious questioning of the validity of a coupon eventually developed into a public humiliation of my mother. It became an interrogation. A gradual destruction of respect. Each coupon she questioned seemed to imply we were dishonest. Each raised eyebrow suggested wrongdoing. It was no longer about saving money on paper. Our integrity was at stake.

And I saw it. I saw my mother—calm, respectful, completely correct—now publicly diminished. I saw a blush rise up her neck. I saw her suppress frustration to keep her composure. I saw the cashier too: her eyes tired and shoulders stiff, possibly taught to prevent loss, possibly assessed on speed and shrinkage rather than on treating people as people. I can picture today, decades later, the pressures she may have felt, the unseen metrics speaking in her head: *Don't miss the loss. Don't make a mistake. Get through the line. Protect the company.* Over time, she stopped seeing the people standing in front of her.

But at six years old, all I understood was that my mother was being treated like a criminal for doing nothing wrong.

Something snapped inside of me. I grabbed my mother's hand and climbed up onto the edge of the bagging area. As my small shoes slid slightly on the metal railing, and with all the righteous indignation a young girl could muster, I declared loudly enough for everyone to hear:

"That's okay, Mom. We'll go someplace that actually cares about our business."

Silence fell across the store. My mother was embarrassed. I took her hand, and we left together through the automatic doors that opened wide with a dramatic whoosh. Outside, I burst into laughter—as did my mom—partly from nervous excitement, partly from victory. Beneath the laughter, however, was something deeper. A line had been crossed. And I knew it.

Remember, I was only six years old. My mother was mortified. That was the day my life's work began.

The Cost You're Paying

I would spend the next thirty years in the very industry that caused moments like the one I had experienced. I built and managed large-scale service operations. I developed policy, technology, scripts, escalation procedures, and automation strategies. I became accountable for the systems that determine whether a cashier, an agent, or a digital interface treats someone with dignity or with suspicion.

I studied why businesses consistently veer away from serving people and toward policing them. Why policy increasingly takes the place of good judgment. Why efficiency gradually replaces compassion. Why trust in the customer erodes into default distrust. I also made my own mistakes, some of which I am still deeply sorry for and which I will share with you in these pages because the wound is what produces the wisdom. You cannot lead people toward something you have not personally navigated.

And I found out what it costs: loyalty, your reputation, your employees' morale, future growth. It costs trust, which is the single currency that makes a brand sustainable over time.

Maybe you're reading this because you already feel it. You're the leader being handed a mandate to cut costs, automate the contact center, and reduce headcount—and something in you keeps asking whether it's the right call. Or you've already made the call and you're living with the fallout. Maybe you've watched your NPS erode quarter after quarter without anyone above you treating it as the emergency it is. Maybe you've sat in a meeting where someone said, "Customers just need to manage their expectations," and you didn't say anything because you weren't sure you had the language to push back.

This book is for that moment and the many others that rise in often well-intentioned pursuits of efficiency yet cost the customer the experience they're entitled to.

The Customer Service Fire

This book is born from that fire—a fire that ignited early and still burns. It is a call to action. Whether you run a small business or a multi-billion-dollar enterprise, this book will help you diagnose what is broken in your service chain, think differently about people, process, and technology, and bring back the caring, human-centered way of leading that your customers and your employees both deserve.

This book is a wake-up call for the small business owner who thinks "good enough" is good enough. For the manager stuck between irate customers and clueless executives. For the newly appointed VP chasing trends instead of fixing the basics. For the CEO who cannot figure out why retention is plummeting despite a bigger-than-ever marketing budget. For any leader who wants to see organizations start caring again.

Come take the journey with me.

The Plan— Reawaken, Rebuild, Reignite, and Reinvent

Before you go any further, I want you to know something: This book ends with a plan.

At the back of these pages, you will find the Beckerman Caring Challenge™—a four-quarter, twelve-month roadmap I developed to translate everything in this book into action. Not inspiration, not principles, but behavioral moves you can put on a calendar and execute, ninety days at a time. The promise is to reawaken what has been lost, rebuild the structures that make caring sustainable, reignite the people around you, and reinvent the way you lead.

The Beckerman Caring Challenge is built on five non-negotiables—five commitments that will quietly test you on every page and measure you every day once you put it down:

Customers are partners, not transactions.
Employees are assets, not expenses.

Trust is capital, not collateral.
Culture is the foundation, not the façade.
Leadership is a service, not a title.

I am telling you this now because it will change how you read everything that follows. You are not working through a diagnosis but methodically building a plan. And the question this plan is designed to help you answer:

What will we be remembered for?
That is not a rhetorical question. It is the work.

MY HUMANITY STANDARD

Throughout this book, I have intentionally selected companies that best illustrate what genuine customer service and experience look like—businesses that respect their employees, lead with empathy, empower their teams, and earn loyalty by consistently putting people first. These companies aren't recognized because they are perfect, but because they prove that caring is possible at scale and that a culture of customer commitment does not have to come at the expense of operational excellence.

I have also intentionally not named the companies that failed—the ones that diminished the dignity of their customers or their employees. This book is not a shame book. It is not a report card of failures or an exercise in finger-pointing. It is a call forward.

The negative stories included here are meant to illuminate systemic issues, not assign blame. Every company referenced—directly or indirectly—has the capacity to do better. Leaders can create better experiences. Systems can be redesigned. Policies can be rewritten. A culture that values efficiency over humanity can be rebuilt, regardless of how far it has drifted.

If you see your company in one of these examples, I don't want you to feel judged. I want you to get curious. Fix what broke. Redesign what no longer works. Protect the humanity in every employee before you strip it away in the name of making things more efficient.

This book is intended to serve as a wake-up call, not a weapon.

Natalie D. Beckerman

Part I–
THE CARING RECESSION

What we previously called a "crisis" regarding "care" suggested an urgent issue to be temporarily addressed—an increase in demand for care, a strained system of care which can recover with sufficient focus and attention. However, the current situation appears to be less of an urgency/short-term need and more of a long-term structural concern. We stopped seeing it as an obligation and started treating it as optional. What follows is the truth about how that happened—and what it's costing you.

We are experiencing a caring recession.

CHAPTER 1:

THE HARD TRUTH

For years, my local dry cleaners worked perfectly well. If I had a question, I could pick up the phone and speak to a human. If I needed an item done differently, I could call them and speak to a human. If I needed something pressed one way, washed another way, or rushed for the next day, a quick thirty-second phone call handled it, or I did during a drop-off in fifteen seconds. Easy-peasy.

Then they upgraded and became a bit more modern. At first, the automation was helpful. I received a text when I dropped off my clothes, and another when they were ready. The texts and the automation supported the experience but didn't replace it.

Then they took it too far and became ultra-modernized. They became fully automated.

One afternoon, I walked in and found no counter, no front staff... no human being at all. Just a locked door and two self-service kiosks. I stood behind another customer, waiting to manually enter every item of my order into a screen. What used to take thirty seconds with a person now took ten minutes of tapping, scrolling, correcting, and hoping I hadn't missed something.

Then I hit the real problem.

The system had quietly removed the $5 next-day service option, the choice I used for years. It simply wasn't there. The options to individually choose how I wanted each item were gone. They made it easy for them, not me. So, I did what any customer would do: I knocked on the locked door. Loudly. Okay, I started banging on the door, not even sure a human was behind it. Eventually, someone appeared.

As I explained my issue, he blankly stared at the screen. "Yeah... it's not there. I'll see if I can get it added back in."

In the meantime, he would make sure my clothes would be available by the next day.

"Tomorrow?" I asked.

Unsure, he told me he would need to make a few calls.

There was no override. No way to confirm. No proof. Just trust.

My automated receipt arrived via text message seconds later, and of course, without an updated delivery date.

The next day I called the store—well, I tried to.

The local number no longer worked. I was routed to a centralized call center that had never seen my clothes and could only tell me what the system said: "Your order is on its way."

On its way from where? When would it arrive? My call was then transferred back to the same dry cleaner—and never answered. This loop continued for over an hour.

Eventually I gave up and drove there myself. When I arrived, the clothes still weren't ready. I waited another thirty minutes. What used to be a thirty-second transaction became a two-hour ordeal.

This is what happens when companies automate without understanding their customer analytics or end-to-end experiences. They didn't test all their use cases or customer scenarios. They did, however, make it easier for customers to pick up their clothes after-hours. I'm not sure if that is their biggest use case. But dropping off dry cleaning is much more difficult than the pickup. You want your clothes done in a certain manner.

What they didn't test was reality.

What happens when a five-year customer needs some items pressed, some starched, some washed, and some returned the next day? What happens when the system removes a service people rely on? What happens when the customer needs help and there is no one left to help? This wasn't a technology failure. It was a leadership decision failure.

They automated a process they did not understand, replaced people before the system could actually serve the customer, and created a maze where a relationship used to exist.

Automation didn't make the experience better. It made the company harder to deal with.

And that is exactly how loyalty dies—not in dramatic moments, but in ordinary transactions that stop working for the human being standing in front of you.

I don't go to that dry cleaner anymore. I take my business to another local shop where I have built a fantastic relationship with the owners and have a laugh each time I go in to drop off or collect my clothes. And I can tell you that I look forward to it.

Now ask yourself the harder question: When is the last time you tried contacting your own company's support line? Did you survive eight menu options, fifteen minutes of hold music, and a monotone agent who couldn't help you? Did your automation work, or did it deflect the problem and make it harder for customers to connect with you?

You know the moment—the exact second a customer realizes the company wasn't built for them anymore. Are you building that moment for someone right now without knowing it?

Businesses keep doing the same thing. A company "jumps the shark" in customer experience when it stops caring about dignity and starts prioritizing speed and cost. This is the moment when efficiency replaces empathy and the brand quietly loses its humanity.

EBITDA, NPS, and business results absolutely matter. Deeply. But not at the expense of caring. That is a false tradeoff that too many leaders have accepted without question. Every organization must find the balance between performance and people, between efficiency and empathy, between growth and humanity.

That is the Caring Recession.

The Caring Recession is a prolonged reduction in emotional, social, and institutional investment into the practice of providing care. The current state of affairs is not solely because care systems are currently unable to provide adequate support; it is also due to the fact that as a society we have de-prioritized, devalued, and continue to erode the ability to provide quality care over time. While a crisis requires some form of response, a recession signifies a continued decline in performance or capacity—one that will shape our future expectations, relationships, and perceived minimum standards of how much care we feel is capable.

And the companies that survive the next decade will be the ones that refuse to "jump the shark."

Now is the moment we must finally stop pretending this is someone else's problem.

It's not marketing's fault. It's not your frontline staff's fault. It's not even the customer's fault for being "difficult." Most customers aren't. Rather, they are tired. They are exhausted from bouncing from associate to associate, agent to agent, channel to channel, chatbots to tickets, and back. They are tired of being treated with suspicion, of believing you are trying to outsmart them, and drained by being asked to do most of the work. Your customer is not the enemy, but they will be if you keep treating them like one.

You see, your internal systems are broken. When service collapses, it rarely does so in one harsh dramatic moment. It instead disintegrates into a thousand small moments when the customer realizes they are on their own.

And if you don't start caring and get it together, your competitor will.

Let's get straight to the point about what's really messed up. Customer service isn't lacking because the folks on the front-line aren't doing a good job. It doesn't work because the leaders set up a system that makes it super tough to provide acceptable service. The processes were set up for your own convenience, not really for the customers.

Your hiring decisions seemed to focus more on filling positions than on actual skills, and leadership experience in service often felt like an afterthought. Tech was brought in to save a few bucks, not to build connections—and honestly, the tech you already shelled out for often gets underused, is badly set up, or leaned on to circumvent issues instead of fixing them.

I can't tell you how many organizations I have worked in that paid millions of dollars and fought for new and better systems and then told the same people in their organization that signed off on these systems just how bad they actually are. This happens repeatedly.

This is why AI isn't going to save you. Not on its own. And letting people go won't help you either. The solution isn't fancy, but it's necessary: You need to revamp the service chain and your poorly designed or absent customer experience path so the business can consistently provide care. We must tackle the three pillars—people, process, and technology—simultaneously since they tend to fall apart together.

When things go wrong, your team ends up being the one to say sorry. When folks feel powerless, tech turns into a barrier. When tech is set up to evade issues, customers end up feeling ignored. These are hard facts. Every leader in your organization has experienced it as a customer. You've experienced the wrath. You've made it through the hold music. You've seen a straightforward problem turn into a weeks-long treasure hunt. That's not the customer just being dramatic. That's the system being broken, not working, or, in all reality, working as designed!

When you fix all three, you suddenly start to realize that your policies are rigid, your approvals chain is too long, and tickets bounce back and forth between departments like hot potatoes. You also finally understand that your people are either empowered or powerless.

Scripts or playbooks don't create great service; people do. Unfortunately, your frontline teams are often underpaid, undertrained, micromanaged, afraid to step outside of policy, and in the dark about how your systems are meant to work. I have watched clever agent after clever agent create workarounds to fix the broken user designs their leaders built. If they follow the system and things go wrong, they are sometimes beaten so badly that it's hard to know what they can do, and they become even more afraid to use their own intuition.

Then leadership acts shocked when service feels cold and mechanical. You want passion? Give your team a reason to care. You want accountability? Start with leadership.

About a year ago, I invested in a high-tech piece of fitness equipment, and within a few months of my purchase, it malfunctioned. It wasn't the reliable high-end product I had been promised. After several futile attempts to resolve the issue with the customer service team, I wrote to the COO. I wasn't angry or dramatic. I simply explained the issue and let him know how much I had invested in their brand. I was feeling incredibly misaligned to the promise they made when they sold me the equipment. To my surprise, the COO responded within hours. His email was polished, empathetic, and reassuring. He apologized. He even used phrases like, "We'll make this right." I felt seen. But his words did not reflect the company's actions, and getting my machine fixed was a nightmare. In the end, I not only had an expensive piece of equipment but an even more expensive realization. Not only do companies lose customers when things go wrong, but they also lose customers when they fail to show up, even after owning it. It's almost like the acknowledgment is good enough rather than putting it right.

Today's customer service is often robotic, reactive, and riddled with red tape. The COO thought his response was good enough, that I would feel happier since he said he would resolve it. But then he passed me right back to the same escalation team I started with.

The reality is many of the people who have contributed to the caring recession are victims of systems that they find themselves in through their nonstop efforts to drive even greater efficiencies across their organizations. The people meant to help your customers are underpaid, undertrained, and burnt out.

Meanwhile, leadership is either out of touch or obsessed with chasing "efficiency," "AI adoption," or the latest tech buzzword without any real strategy. I had a leader tell me, "We need to use agentic AI." When I asked why, the answer was, "To reduce staff." There was no vision. No problem statement. No plan to improve customer experience. We'd already cut 25 percent of the team based on requirements. The irony:

We were a service company, yet nothing about our recent decisions was improving our service.

Let's be clear: This isn't just a "service" issue. It's a systemic failure of your broken processes, mismanaged people, and your misuse or nonuse of your technology. Fixing this isn't optional anymore. It's your only option. And it needs to be your number one priority.

Whether you have two, twenty, two hundred, or two hundred thousand employees, the mission is the same:

You must make customers feel seen and heard. You need to help them solve their issues quickly while you treat them as if they truly matter. Your job is to make it easier for them to do business with you and to show your customers you really care.

This book will show you how to build a service culture that works, from hiring and onboarding to development and recognition. Let's keep going.

CHAPTER 2:

YOU'RE NOT THAT SPECIAL

One of the largest luxury sportswear brands claims to believe their lifestyle collection is the epitome of modern wellness. The brand created such an exclusive image of itself that when you walk into the store, you feel as if you have entered a temple of carefully curated aspiration. The lighting is perfectly arranged. The merchandise is showcased as if it is artwork. The color and design of the shelves and racks are inviting and eye-catching. The exorbitant prices of the merchandise create an aura of privilege. The brand believes it is completely distinctive.

However, when I tried to obtain assistance, my experience told a different story. Whenever I contacted them to report an issue related to my account—a relatively simple matter—the brand treated me to a frustrating cycle of waiting to speak to someone and/or calling back and waiting for a representative to contact me. Each time, I was subjected to the same aggravating process of calling back, waiting again, and describing the issue to another representative who seemed to not understand the problem and respond to me defensively, dismissively, and oddly, proudly.

Although the in-store experience of the brand created an atmosphere of "luxury," the service behind the scenes told a different story: "You're on your own, good luck." The response to each of my complaints reflected its disconnect from reality: The brand was not unique; it was simply disconnected and relying solely on their own mythology to define themselves as superior.

In choosing to prioritize superiority over resolving a basic account problem, I realized the brand's efforts in being unparalleled would collapse rapidly when they failed to resolve the issue or follow-through on fixing the problem. No brand is unique based on their claim to uniqueness or their products and features. Only those that continuously focus on valuing, respecting, and seeing their customers' needs will earn the right to that singular reputation.

The unpleasant reality most leaders do not want to hear is that your product is no longer a defense mechanism. Your product's features are no longer a barrier. Your shiny new logo is no longer a differentiator. In today's competitive marketplace, competitors can easily replicate everything you build before you can recover from one negative customer experience. However, what they cannot replicate is how you treat customers. Or how easy it is to buy from you. Or how quickly you can correct problems when something goes wrong. Or how you respond when customers are frustrated, anxious, or disappointed.

These are the reasons the brands that ultimately win are not necessarily the ones that have the most impressive features. Rather, they are the easiest to buy from, the most human, and the most consistent. Brands that respect customers' time. Brands that do not penalize customers with excessive process. Brands that do not use policy as a tool against customers. Brands that eliminate friction wherever possible because in a competitive marketplace, they can ill afford to do otherwise.

Removing friction for the customer and within the workplace is the most under-rated solution available to solve customer experience problems.

Let's pull the band-aid off: Your products and features are no longer a defense mechanism.

Your app or shiny new logo is no longer sufficient to differentiate yourself from your competitors. They can and will replicate anything you build virtually overnight. How easy are you to buy from? How quickly can you fix a problem when something goes wrong? How do you choose to show up to customers every day? Winners are not always the "best" on paper. Winners are the easiest, most human, and most responsive to the needs of their customers.

You may sell a product or service that customers can find elsewhere. What customers cannot get somewhere else is your experience. If your return policy is a scavenger hunt, your chatbot is a maze, and your representatives seem to prefer to be anywhere else, your customers will abandon you in an instant.

Customer experience is the real differentiator, and your policies will either establish trust or destroy it. When you prioritize ease, speed, empathy, and humanity in every interaction, customers will trust you and remain loyal. However, if you continue to make returns difficult, utilize chatbots that are impersonal, and force your representatives to read from scripts, customers will flee to your competitors.

Being "the best" is not necessary. Being easy to work with and responsive to the needs of your customers is.

Two comeback stories clearly illustrate this point. One skincare brand had incredible packaging, innovative marketing, and social media buzz, but when the product failed to meet expectations, the return process was akin to trying to navigate a confusing maze. The customer had to call during specific times, wait on hold, receive a code via voicemail, submit a receipt electronically, wait several days for approval, and pay to ship the item back. The product was not the issue, the experience was.

A competing brand took a different approach: an easy-to-use web portal that generated a return shipping label instantly. Once the shipping carrier scanned the package, the competing brand processed the refund. Then, it followed up with a friendly message: "We apologize it did not work out for you. Please come back when you're ready." Both brands existed in the same space and the same categories, but a completely different level of respect for the customer existed.

The theme of this chapter is singular: You can't assume loyalty simply because you want it. You must earn loyalty each day by being easy to interact with, responsive, and relatable, particularly when something goes terribly wrong.

Creating obstacles for customers to conduct business with your organization, hiding in a sea of fine print, requiring customers to jump through hoops to resolve simple issues, and believing your product is so excellent that customers will tolerate it forever, will never work.

You must continually strive to remove friction. Be obsessed with ease, speed, and consistency. Give your team members permission to behave as humans, not policy drones. Earn loyalty by making a conscious effort to do so every single day. Customers always have a choice, and they will exercise it the first time you make things difficult.

If you are too difficult to do business with, you are speeding along the road to ruin. Let's look at some examples of organizations that understood this vs. organizations that did not. Many organizations ultimately failed because of their complete indifference to the wants and needs of their customers.

Peloton vs. Apple Fitness+

Peloton built one of the most passionate customer communities in modern consumer history. Their early riders didn't just use the product—they evangelized it. Membership grew to nearly three million connected subscribers.[1] Instructors became celebrities. The brand felt like a movement.

Then leadership stopped listening. When the Consumer Product Safety Commission flagged their Tread+ treadmill after injuries and a child's death, Peloton fought the recall rather than owning it. When pandemic demand cooled and customers signaled, they were returning to gyms, leadership doubled down on factory expansion instead. When the product line got confusing and subscription prices rose without new value, churn climbed. By early 2022, Peloton had halted production with 500 days of unsold inventory in warehouses. The stock lost more than 90 percent of its value from its peak.

Apple Fitness+ launched in the middle of all this—no hardware required; available on devices customers already owned; and $9.99 a month. Apple didn't beat Peloton with a better bike but instead by removing every barrier between a customer and a workout. Peloton had the community. Apple had the friction-free experience. The customers chose ease.

Meta vs. TikTok

In 2021, Mark Zuckerberg renamed his company Meta and announced that the future was the Metaverse—a virtual world where people would work, socialize, and shop through virtual reality headsets. Over the next three years, Meta invested more than $45 billion in this vision. At its low point, their Horizon Worlds platform had fewer daily active users than some mid-sized city Facebook groups. The avatars had no legs. Nobody showed up.

Meanwhile, TikTok was obsessively watching what users actually did—how long they paused, what they rewatched, when they scrolled away—and used every signal to serve the next video. The algorithm didn't predict what customers wanted. It learned it, in real time, from behavior. TikTok reached one billion users faster than any platform in history.

Meta only stopped the bleeding when they copied TikTok's format with Reels. The company that declared itself the architect of the internet's future saved its advertising business by following the company that simply refused to stop listening to customers. No product, no brand, no legacy insulates you from that lesson.

Failing to identify the direction in which the marketplace is moving, and not prioritizing the wants and needs of customers today or tomorrow, will ultimately lead to more retailers going bankrupt, even though they could have continued to evolve by trusting their customers, honoring their wishes, and recognizing that the competition is also listening and evolving faster.

Toys "R" Us vs. Amazon

Then, there are the companies that create an environment where the customers can transact easily and efficiently. These companies create a customer-centric culture. What a great lesson from what could have been. Who didn't grow up listening to the Toys "R" Us jingle? "I don't wanna grow up, I'm a Toys 'R' Us kid."

Toys "R" Us had physical stores, where customers could browse the merchandise. Amazon did not. Toys "R" Us continued to focus on the wrong aspect of their business... price instead of the experience. With all its space, Toys "R" Us could have crushed their competition with birthday parties, events, gift wrapping, and so much more. But, Amazon,

Walmart, and Target offered low-cost, aggressive online shipping and priced products competitively.

Although Amazon played a role in the decline of Toys "R" Us, which started in 2000, it had a ten-year partnership to sell the company's toys and baby products online in exchange for being its exclusive supplier.

Toys "R" Us agreed to redirect all its online traffic to Amazon. By 2002, the e-commerce partnership—which cost Toys "R" Us $50 million annually—had exceeded all of Amazon's expectations—and not the best decision Toys "R" Us made.

Once Amazon allowed other merchants to sell toys and baby supplies on their site, Toys "R" Us sued for breach of contract. After winning the $51 million lawsuit, Toys "R" Us terminated the contract. Many are unaware of this dirty little secret: The original intent of the agreement was never to increase Toys "R" Us's and Amazon's combined toy sales, but to eliminate the competition. Sneaky, huh? Amazon wants to own a market segment, rather than compete within it. And it used the largest toy retailer in America to convince customers to buy online, rather than during a visit to a store.

While using the Toys "R" Us brand to help them achieve this goal, Amazon simultaneously prevented the company from developing its own online shopping experience and placed it years behind in development and innovation.

Despite its decades of success and dominance in the industry, Toys "R" Us wasn't aware of the threat to business from internet shopping, and specifically Amazon. With a lack of profit diversity, Toys "R" Us grew vulnerable to the threats posed by Amazon.

Leveraging a strategic first-mover advantage, Amazon carefully planned and executed its approach on Toys "R" Us, waiting until the company's business model had deteriorated enough before finally ultimately driving it out of business.

Amazon had the financial resources to sustain short-term losses and achieve its long-term goal of eliminating the competition, while also wielding the foresight to realize how quickly the battlefield was changing. That's the power of this story. Amazon absorbed a short-term loss for long-term gains AND risked where the market was heading. Toys "R" Us was woefully behind in realizing its own vision and slogan.

Amazon remains comfortable in maniacally providing a consistent experience for its customers: one-click buying, continuous improvements, painless returns, and continuous feedback loops. There are ginormous lessons to be learned here. You are not entitled to loyalty. You must earn it, every single day. If you quit caring about your customers and their needs and preferences, then you will lose every time. Goodbye for good.

FROM EXPECTATIONS TO EXECUTION: WHERE BUSINESSES WIN OR LOSE

Businesses that truly stand out either win or lose by transforming expectations into execution. They win or lose in the gap between the two.

By now, the pattern should be crystal clear. Customers do not remain loyal to you because you tell them you care. They remain loyal to you because you demonstrate that you care in every interaction. Features can be replicated. Marketing efforts can be dismissed. Products will eventually be matched.

But what customers judge is whether you make their lives easier, whether you show that you value their time, and whether you keep your commitments without forcing them to battle your systems to get to those commitments. This is where the real game is played.

We can draw a straight line between what customers expect, the levers that leaders have at their disposal, and the areas in which businesses quietly fail when their processes, people, or technology are constructed for internal convenience, not for humans.

Repeatedly, companies face the same choices: Invest in experience and build trust or ignore it and slowly lose relevance.

Here's the part of the message that most leaders do not want to hear: Customers do not evaluate you based on your intentions. They evaluate you based on your results and your recognition of their needs.

Customers do not care about how many times you have met to develop your strategy, how many millions of dollars you have spent to rebrand, or how difficult it is for your department heads to align with each other. They care about whether you resolve their problems, whether they can access you through the communication method that fits their lifestyle, and whether you treat them like a person instead of a transaction.

And the bar has been raised and continues to rise.

Customers are no longer evaluating you solely against your competitors. They are evaluating you against the best experience they ever had—whether that was Amazon's one-click checkout, Chick-fil-A's consistently friendly service, or Apple's seamless in-store support. Once people experience that type of ease and respect, they do not unlearn it. They do not lower their standards simply because your internal processes make it difficult for you to resolve their problems.

At the heart of this transition are five expectations that have become non-negotiable. Customers expect to receive prompt, useful responses to their inquiries—not just fast for the sake of being fast, but solutions to the problems they are experiencing. Customers expect to interact with you however they wish to, on whatever channel is convenient for them in that moment. They expect to carry their story with them wherever they go and not have to begin anew with you each time. They expect reasonable policies that benefit people, not just protect the interests of your company. And when things inevitably go wrong, customers expect empathy, ownership, transparency, and follow-through.

Rarely do customers abandon ship because of one bad experience. They leave because of the cumulative effect of unmet expectations. If you miss even one of these expectations repeatedly, they've already begun shopping elsewhere. If you miss two of these expectations consistently, you've lost them. Meeting these expectations gives you a chance to survive. But surviving is not winning.

The businesses that win understand that experience is not a slogan—it is an operational discipline. And it is fueled by three factors that influence every customer's decisions: effort, emotion, and trust.

Effort is the amount of work you require your customer to complete. Every unnecessary click, transfer, policy barrier, or repeated explanation erodes their loyalty. The less work you make for your customers, the more likely they are to remain loyal.

Many of us despise the United States healthcare system. All the effort is on the patient (but that will be for another book). We have one of the most painful healthcare systems in the world, and it is getting worse. The customer has to actually make the effort to get things done, the system does not assist the customer in any way with appointments, follow-ups, scheduling, and the list goes on and on.

Emotion is how the customer feels once the interaction is completed. People may forget specifics, but we all know they never forget how you treated them emotionally. One instance of genuinely empathizing with your customer can counteract a single failure. Conversely, repeated instances of ignoring the emotional needs of your customers will undermine even the greatest product.

Trust is whether you uphold your commitments. Customers forgive errors when they perceive you to be honest, transparent, and committed to fixing things correctly. When you break that trust, the relationship does not blow up, it simply evaporates.

When you consistently deliver low effort, positive emotions, and high trust, you become the default choice—whether your product is perfect; whether your competitors are cheaper; whether things go wrong. Customers will excuse imperfections in your product if the way you sell, serve, and support them is easy, humane, and reliable.

This is where businesses either differentiate themselves from the crowd or quietly fade into the background.

This chapter destroys one of the most hazardous myths in modern business: the idea that you are invulnerable due to a strong brand, a remarkable product, or a devoted customer base. You are not.

No business is "unique" due to its hype, heritage, pricing, or advertising. Only when businesses continually improve the lives of those who pay them can they be considered extraordinary. Although the high-end sportswear brand in this chapter appeared flawless on the surface, much of what it sold were aspirational qualities. Confidence radiated from the brand, and its mythology was part of the appeal. However, once the brand stopped addressing real customer needs, the illusion disappeared.

Peloton, Meta, and Toys "R" Us were all, at one point, dominant, and believed they would be protected from competition by a loyal following, their product quality, or their sheer scale. Each failed to take into consideration how rapidly friction—or indifference to what customers actually wanted—would cause customers to flee.

Customers do not leave because you are not the best, as this chapter makes irrefutable. They leave because you are difficult to deal with, difficult to locate, difficult to understand, difficult to contact for answers, and difficult to trust.

Being difficult is a death knell in a world where options are merely a mouse-click away. The most innovative or flashy businesses don't always endure. Those who prioritize consistency, speed, and simplicity remove friction, empower people to resolve issues, and avoid letting ego, process, or policy penalize customers. You do not deserve loyalty. You must earn it every single day.

This chapter illustrates through personal experience and real-world examples that branding can be overlooked, features can be emulated, and products can be replicated. But the way you treat people when things go awry is something that cannot be reproduced. You are not competing based on what you sell, but on how doing business with you feels. Make it simple. Give it a human touch. Make it dependable. Or watch your customers find someone else who will. Now is the moment to care more—and care more deeply.

CHAPTER 3:

UPGRADE THE GOLDEN RULE TO THE PLATINUM RULE

From an early age, I developed an extraordinary capacity to shop. I am both a serious and a seasoned shopper—usually on a mission and with clear objectives and goals; I do not just buy things for fun. When I give someone my last dollar, I do so with expectations, both from the purchase itself, and from how I am treated while purchasing it.

Eventually, I realized there is one thing that most businesses still fail to recognize: That caring is not just a personality trait, but a very smart business model.

The Golden Rule, which is simply "do unto others as you would have them do unto you," has been a guiding force for thousands of years. Today, it has evolved to "Treat others the way you want to be treated."

It is part of many cultures and religions; from the teachings of Confucius in ancient China, to the wisdom of Jesus, to the laws of high-order rabbis and religious scholars throughout history. It became the ethical benchmark of civilization, a simple yet universal baseline for ethics. But today, the Golden Rule is no longer exceptional. In fact, it is the norm in many businesses, yet in others, it's barely alive. And I would say, it is no longer good enough. Many customers and employees now demand the **Platinum Rule: "Treat people the way they want to be treated."**

That is a huge shift. It is not just semantic but structural. The Platinum Rule is a fundamentally different philosophy. It requires flexibility, emotional intelligence, cultural sensitivity, compassion, and courageous leadership. The Platinum Rule dictates and demands that organizations should move away from building around policies and build and design around people. It is a governing principle for organizational culture, customer experience, employee engagement, and psychological safety;

the intangible infrastructure that determines whether people feel safe, valued, and respected within your organization.

In my early career, I experienced the lack of that principle firsthand during what should have been a very simple transaction. Office supplies are my favorite thing to shop for, outside of clothing. Walking into an office supply store always brings me great joy and excitement, especially in the notebooks and pens section. There is something about a new notebook... it's not just a fresh start, but rather the first step of a new, exotic journey. I should have been the easiest customer to retain. Instead, I became a case study of how to lose customer loyalty to your competition.

During my early years of consulting work, I spent an extraordinary amount of time in my local office supply store. However, that store began to establish itself as a place that viewed customers as disruptions to their rules rather than customers who enabled the business to grow. I sensed that.

On my regular Friday evening visit to plan for the upcoming week, I was greeted by rigid policies and procedural scripts while attempting to return an item or inquire about pricing. Rather than ask how they could assist me since I was a frequent patron, I was met with, "It's past fourteen days," or "You don't have a receipt." That became the standard. It played out like a blockbuster movie in "policy management."

This company used policies as an excuse to avoid being empathetic to my needs. They debated me on EVERYTHING. The brand had everything necessary to create a loyal small business in a local community and dominate an important category—office supplies.

However, they developed an organizational model built around rules and policies rather than people. They didn't ask how they could fix the issue or assist me, but rather, *how can I push you toward the competitor?*

They ultimately did—to Amazon, Walmart, and Costco. Treating customers as they want to be treated, not as dictated by a policy manual, is not revolutionary. Neither is The Platinum Rule. But it is about survival.

You need the Platinum Rule because it is based on tangible business differentiators, as well as empathy and common sense. The Platinum Rule is about recognizing customer needs and responding to those needs. As a result of the digital economy, and the availability of alternatives only one click away, businesses must evolve. They must learn how customers and employees want to be treated, and then design their systems, training, and leadership behaviors to accommodate those truths.

This is not about being nice—this is about being successful. Organizations that win today are the ones that develop the foundation of their organizational models using dignity, care, and adaptability. When customers feel heard and understood, they remain loyal. Conversely, when customers are perceived as unimportant or are drained by interactions with a business, they leave—often quietly—permanently and with their network.

You'd think this would be obvious. Yet in far too many organizations, genuine service gets buried beneath protocols, rigid policies, and the relentless pressure of quarterly objectives. At some point, somewhere, the fundamental principle of human interaction, empathy, loses traction.

Empathy doesn't require dramatic gestures. Empathy requires leaders who choose dignity over defensiveness and whose systems empower employees to do the right thing easily. This is why the stories of organizations such as Chewy, Nordstrom, The Ritz Carlton, Delta Airlines, and Trader Joe's resonate so broadly. Those stories convey a signal that people are crying out for: "You Matter Here."

The Platinum Rule is what turns that signal into culture. The Platinum Rule produces organizations that foster an environment in which employees feel psychologically safe to act ethically. Organizations that support values that are not only articulated in writing on walls, but also in action. Organizations that understand that trust is not merely an aspirational goal, but a daily practice. Organizations that recognize that kindness and respect are inexpensive. Indifference is an expensive proposition.

Empathy builds a reputation faster than marketing ever will. Common sense generates uncommon loyalty. Caring is far from outdated—it is today's most powerful competitive advantage in business.

When you design and develop your organization around these ideas, something remarkable occurs: Customers become more forgiving. They remain loyal longer. They frequently advocate on behalf of your organization. That is not sentiment. That is solid business.

There is a key reason why more organizations don't adopt this approach: fear. Fear of abuse. Fear of inconsistency. Fear that empathy will not scale. However, the larger truth is this: You can scale empathy if you design for it. You can construct guardrails that allow judgment. You can train employees to own issues, and you can reward them for resolving issues, not deflecting them.

This chapter presents the opportunity to reclaim the baseline: Treat customers as they wish to be treated, not per your policy manual. It is a straightforward concept. It is also the most underutilized method for developing growth strategies in contemporary business.

It is about being effective. In a digital economy where each alternative is only one click away, what you are actually offering is not merely your product or service, but the manner in which you make individuals feel. When customers feel recognized and understood, they stay—otherwise, they leave, typically before you realize it.

Common sense and empathy are not luxuries. These are competitive weapons. In an era in which alternatives are only one tap away, your actual product is the experience of interacting with you.

WHAT EMPATHY LOOKS LIKE—AND WHAT IT DOESN'T

Empathy is the most misunderstood driving force in business. Yet it's easy to dismiss it as "too expensive" or "unscalable." The reality is the finest illustrations of empathy in business are seldom costly or complex. They are rooted in something more substantial: common sense and elementary human decency.

But that isn't true or accurate. Empathy is not dependent on elaborate programs or enormous budgets. Empathy is dependent upon leaders who choose to view the individual standing before them as a human being, not a problem.

Consider what that looks like when executed properly. In the aftermath of a weather-related disaster, an airline gate attendant was confronted with a decimated schedule. While she was unable to alter the storm, she could modify the experience. She provided snacks to stranded passengers, reported accurate updates every fifteen minutes, and assisted elderly travelers in rebooking their flights. There was no pre-determined script or prior approval from corporate. She simply chose to perform the obvious right thing, and her actions created memories that endured long after the storm subsided. People remembered that moment long after the storm passed, and they told others.

Now compare that with what happens when empathy ceases to exist.

When a customer's vehicle was totaled, their large insurance company informed them a rental vehicle would be available within twenty-four hours. But five days and fourteen telephone calls later, the rental vehicle had not arrived. The customer was transferred between departments, countries, and representatives, who were all equally helpless and equally disengaged from the initial commitment. The story made headlines. Not because of the vehicle loss, but because this large insurance company's indifference was unacceptable.

Empathy manifests itself in less prominent, more consistent ways as well.

At Nordstrom, empathy and customer-centric service are not slogans. They represent operational principles. You may recall the Nordstrom tire story, a well-known anecdote from 1975 in which an employee in Alaska fulfilled a customer's request to return a set of tires (a product Nordstrom never sold), and in the process exemplified the company's motto, "Use Best Judgment in All Situations." By prioritizing the desires and trust of the customer above established policies and procedures, the employee exhibited profound empathy and converted an unreasonable request into a distinctive brand legend.

As the story goes, a customer was adamant he purchased the tires at the Fairbanks store and was therefore entitled to return them at any time. But Nordstrom didn't sell tires. However, it had acquired three stores from Northern Commercial of Alaska in 1975, which sold a vast array of products, including linens, towels, and automotive equipment. When Nordstrom assumed ownership of the stores, the merchandise assortment was limited to clothing and footwear. The employee wished to provide fairness to the customer, who had traveled approximately fifty miles to return the tires, rather than discard them. Unfamiliar with the price of tires, the employee contacted a tire dealership to determine the estimated value of the set, then subsequently accepted the tires and provided the customer with a refund.

Unfortunately, the exact amount the customer received in exchange for those tires has faded from memory over time. Yet it is likely that that value—both in scale and significance—has been more than recouped many times over. The tire story has become such a pivotal component of Nordstrom culture that it continues to serve as a living and authentic example of the company's core service-based values. It provides employees with the authority to use reasonable discretion to satisfy customers, and it serves as a powerful enduring example of how a company can differentiate itself through superior service. It is a symbolic representation of the company's culture and a constant reminder that employees are empowered to act responsibly without requiring manager approval.

Why I love this story so much is that even 50 years later, it serves as a living, authentic example of Nordstrom's core service-driven values. It empowers employees and authorizes their staff to make reasonable decisions to satisfy customers, and it serves as a powerful enduring example of how to differentiate a brand through exceptional service. It is a cultural symbol and constant reminder that employees are trusted to do the right thing without needing manager approval.

This concept of trust, empathy, and care for others' welfare is evident elsewhere.

When a customer contacts Chewy to cancel an auto-shipment because their beloved pet has passed away, Chewy doesn't simply process the cancellation. They send handwritten condolence cards. Flowers. Sometimes a custom portrait of the pet, painted from a photo.

Customers shared these experiences online. The posts went viral. The earned media generated by those simple acts of human decency far exceeded what any marketing budget could have replicated. Chewy's name became synonymous not with selling pet food, but with caring about the people who buy it. That is the Platinum Rule in action. You didn't ask what the policy was. You asked what the human being in front of you needed.

The expense of doing so is nominal. The impact is lasting: *We care about you, and we are empathetic.*

Patagonia has built an entire philosophy around the same principle—restraint as empathy. Rather than encouraging customers to replace worn items, Patagonia actively recommends they be repaired first. That is a direct contradiction of short-term sales logic. But it has created extraordinary long-term loyalty among customers who trust that the brand shares their values. Empathy does not always take the form of giving something. Sometimes it takes the form of holding back.

LEGO quietly replaces lost pieces—often for free and with remarkable speed—for children who write in to say a brick has gone missing. Parents share these experiences on social media with the kind of genuine warmth that no advertising campaign can manufacture. The result is multi-generational loyalty: Children who grow up knowing LEGO cared become adults who buy LEGO for their own children. That kind of loyalty cannot be purchased. It can only be earned, one small act at a time.

At The Ritz-Carlton, every employee has the authorization to spend up to $2,000 to resolve a customer's issue on the spot. It is not because they desire to give money away, but because their leadership recognizes that establishing trust in the moment will protect the company for years to come.

At Trader Joe's, when a snowstorm prevented a family from grocery shopping, an employee personally transported groceries to the family's home. There was no procedure in place for that—only humanism. Customers do not forget experiences like this.

Why don't so many more companies act similarly? Perhaps they fear being taken advantage of. Maybe it's just a simple lack of care. The bottom line is kindness and respect aren't expensive, but indifference is very costly. Empathy builds brand value quicker than the best advertising ever has. Common sense builds uncommon loyalty. Being caring is not old-fashioned; it is the most powerful competitive advantage that exists today.

When a company designs itself on these core values, something amazing occurs. Customers are more willing to forgive, customers stay longer, and customers will speak out more loudly on behalf of a company that has shown empathy. This is not feeling or emotion; this is a business reality.

The companies in this chapter—Chewy, Patagonia, LEGO, Ritz-Carlton, Trader Joe's—did not win loyalty through grand gestures. They won it through small, specific decisions to treat the person in front of them as a human being rather than a transaction. None of it required a budget meeting. Most of it required nothing more than permission.

So here is the question worth considering before moving on: Where is the one place in your organization right now that a customer or an employee is being treated as a transaction instead of a person—and what would it take to change that this week, not this quarter?

CHAPTER 4:

EVERY TOUCH-POINT IS A TESTIMONY, GOOD OR BAD

People often say to me, "You walk with a sense of purpose."

I am a faster-than-average walker and usually have a specific destination or vision in mind when I'm shopping (we already established that I'm an intentional shopper). And because I have hated grocery shopping since I was a kid, my absolute favorite thing is to walk into a grocery store, ask one of the employees where a product is located, and hear them say, "Let me show you," or "Let me take you to that aisle." That little touch point and assistance has a far greater effect on me than the employee realizes.

Why? Because it saves me time. It shows me I am valued.

Walking into a store, opening an application, calling a contact center, or hearing the first few seconds of hold music, all create a touch-point between your brand and your customer. They create opportunities for your customer to develop a positive relationship with your brand (loyalty) or develop a negative relationship with your brand (destroy).

Each time your brand touches one of your customers it will be a testament to how you care for that customer. There will be those that follow the path of BlackBerry; there will be those that follow the path of The Ritz-Carlton. I have had fantastic experiences with Nordstrom at every touch-point, from the initial call to request a product check and price match to the final touch-point at the desk. I developed a testimony for good. The example of the tires at Nordstrom built brand belief and trust instantly and developed a fifty-year testimonial for that store. It also formed first impressions and aided in the development of the brand's overall reputation for many years.

There are too many businesses that view customer service as an isolated function of a department, rather than part of the DNA or bloodstream of the entire organization. Customer service is present in the load speed of your website. It's shown in the visibility of contact options, or lack thereof; in the tone of the automated messages and auto-reply emails; in how employees speak when they're under stress; in how quickly a simple issue becomes complex.

Recently, I contacted a company regarding a very expensive, nearly new product that broke. I believed that the product breaking on its third use was not due to normal wear and tear—and I paid more than I should have for that to be my experience. The company informed me they would send a replacement immediately. I was blown away. I thought, wow, they must feel the same way I do about quality. They truly stood behind their brand—I was prepared to argue and fight.

Once the representative confirmed—after I had already provided photos—that a replacement would be sent, I received an email requesting more information with a note that it was actually not covered. When I called back, I was transferred to text messaging to finalize the conversation. I was extremely upset by this decision. In the first text message, the representative said they were doing me a favor and noted that the product was nearly a year old. I was furious. I already provided photos, and I had been told a replacement would be issued—they had seen the damage, yet somehow it was being framed as a favor? I repeatedly asked to speak with a live person and was eventually able to reach an agent the next day. I was very confused—what had started as a great customer service experience had quickly turned into a poor one because of a single email, a single text message, and one interaction.

In the end, the company did replace the product; however, it took numerous weeks and interactions with multiple representatives. Every one of those interactions contributed to that brand's reputation and my testimonial. They could have had a great testimony from me.

As we discussed earlier, every touch point between your brand and your customer is an opportunity to develop a positive relationship with your customer and/or an opportunity to develop a negative one. Regardless of which option occurs, the customer will develop a testimonial, which will contribute to your brand's overall reputation and perceived value.

Customers do not experience your company through individual "channels." They interact with your company as one brand, one relationship, one promise. When the website states one thing, and the call center states another, the customer does not think, *Oh, that's a systems integration issue.* The customer thinks, *This company can't be trusted.*

Therefore, the seemingly "small" moments matter more than leaders realize. The sale is not the entire experience. It is the experiences in between the sale: time to answer, ease to return, respect in the tone, clarity of the policy, and the ability to reach a human when the issue is emotionally charged.

One of the greatest mistakes businesses make is believing that only the *large moments matter*—the sale, the refund, the VIP meeting, the "we're-so-sorry" escalation. However, this is not how customers perceive your company. The opinions of customers are formed in the small moments that either reassure or irritate.

Consider the following questions: Was your website easy to navigate? Was the response to that email timely? Does your app freeze or crash frequently? Is the tone of the automated message/automatic reply consistent with your brand's personality? Do your employees demonstrate a caring attitude while dealing with a stressed-out customer or is it obvious the customer is simply a ticket number in a queue?

Trust is generally built in small ways over time and destroyed in a single instance.

The following examples illustrate how trust is built and destroyed across industries—and why the difference between the two is rarely about the mistake itself.

For decades, a major automobile company positioned itself as a symbol of reliable German engineering. Then, in 2015, it was discovered that they had installed software into their vehicles to deceive emissions testing. Customers didn't just feel let down. They felt deliberately deceived. In a matter of days, decades of brand equity evaporated. Sales plummeted. Billions of dollars were paid out in fines. The leadership team was dismantled. The damage was not caused by an accident or a product failure. It was caused by a deliberate act of dishonesty. Trust is not broken by mistakes. It is broken by dishonesty.

A leading American multinational financial services company promoted itself for years as a relationship-focused bank. It was later discovered that employees had opened millions of unauthorized accounts to meet aggressive internal sales targets. Customers felt their privacy had been violated. Employees felt they had been forced into unethical behavior to survive. Initially, leadership attempted to deflect responsibility. The trust collapsed rapidly—not because of a single incident, but because the culture had been quietly destroying trust long before it made headlines. A culture that rewards the wrong behaviors will always produce the wrong results, whether anyone is watching.

A major airline had invested billions in loyalty programs and brand recognition when, in 2017, a single video changed everything. A passenger was physically dragged off an overbooked flight at Chicago O'Hare while other passengers filmed in disbelief. While the incident was devastating enough, the airline's response—blaming the passenger, citing policy—made it catastrophic. The company's market value fell by nearly one billion dollars in the first two weeks after the event. When customers feel dehumanized, trust doesn't erode gradually. It collapses instantly.

A major pharmaceutical company's handling of the 1982 Tylenol tampering crisis remains a reference point for crisis leadership done right—a company that chose long-term trust over short-term profit.

The lesson across all four examples is consistent: Trust is created through thousands of small interactions and protected through the decisions leaders make when it is most costly to do the right thing.

In many cases, a customer's first impression of your service will not require a human to provide the service. It may be the design of your website. If a customer is unable to locate what they are seeking, including your contact information, or is unable to complete a purchase or view the mobile version without experiencing extreme frustration, you are not creating a design problem—you are creating a trust problem and will likely lose the customer.

Swipe up! Bad design equals friction. Friction equals frustration. Frustration quietly destroys loyalty.

A large furniture retailer developed an aesthetically pleasing brand identity using gorgeous product images and clean design. Unfortunately, the retailer failed to develop a clear and concise communication of shipping timelines. Customers purchased $1,000-plus in products, anticipating delivery in approximately one week. Unfortunately, the actual delivery time was six to eight weeks. Predictably, angry calls flooded in, customers canceled purchases, and negative reviews saturated the internet. The failure to communicate a critical aspect of the buying process resulted in a loss of millions of dollars in sales, as well as a long-term decline in brand equity.

Your website is not simply marketing. It is customer service in smaller pieces. Providing transparency and providing ease are your best tools to determine your survival. Your website is merely the starting point.

There are two more things that really get under my skin and make it hard to believe your brand actually cares. I use third-party delivery platforms and food apps regularly. When something arrives with an issue, you push me to your website, but I cannot complete the transaction, so you force me to call you. Then, when I call, you either rectify the situation or add injury to insult: You transfer me to another department or team that is closed. On one recent occasion, I was redirected to a VIP team and placed on hold for—not exaggerating—two hours. I finally hung up and realized they couldn't care less.

I also don't understand that if a product arrives with an issue, I can't reach your company at all—phone, email, text, Instagram, nothing. How it that even possible today? Yet it is. Too many companies act like once they've made the sale, that's it. In reality, it's the exact opposite. I've had enough, and now I'm putting my energy into telling others about my experience.

I recently spent nearly $700 taking part in more online shopping. When my items finally came, they were stuffed in a box in a way that looked like a five-year-old wrapped them. Tape everywhere, no tissue paper on my folded clothes... my heart sank. How could a professional do this?

A reputable brand? I took pictures and sent a message that relayed my sentiments: A toddler could have put more effort into this and certainly would have cared more. Every touch-point matters.

Email Response Times: Your Silence is Deafening

When customers extend your company trust and attention by contacting you through an email or contact form, they expect a timely response. If they receive a canned response, or a robotic, "We'll respond in two to three business days," or worse yet, no response at all, you indicate that you place an incredibly low worth on their time. And in fact, your no-reply email told them exactly how much you value their time: not at all. I am still amazed at the prevalence today that any company would offer a two- to-three-day response time. They may as well just say instead, "Take your business somewhere else." For me, any delay beyond instantaneous acknowledgment is a sign of indifference. Customers expect receipt of their inquiry and/or resolution to occur in real-time.

Use email as a follow up but not to transact. It's too slow.

Travel Agency Experience

A small, boutique travel agency understood the importance of rapid response. During peak travel seasons, the agency responded to inquiries within *fifteen minutes*, often to simply indicate that "*We're looking into this and will get back to you shortly.*" This rapid response created an expectation that the larger travel agencies could not meet. The result? High repeat business and referrals simply because the agency showed up.

If you are unable to meet the rapidly evolving expectations of modern customers, please consider removing email as a primary means of customer service. Please do not continue to hide behind the "contact us" page. Consider providing customers with alternatives to contact you in real-time, such as chat, text messaging, or calling.

The Public Stage You Can't Ignore

Social media is no longer simply a marketing tool. Social media is a customer service platform, a public stage for the world to see.

Your customers are interacting with you on social media platforms such as Instagram, TikTok, LinkedIn, Twitter (yes, we're still calling it that), and more. If you don't respond or worse, respond defensively, you're not simply losing one customer, but you are creating a public warning for all other customers.

Wendy's developed a cult-like following by being responsive, human, and platform relevant. Wendy's responded to social media inquiries quickly, humorously, and in a manner that reflected the personality of the platform. Wendy's social media team did not simply answer questions; they developed a personality for the brand in real-time.

Unfortunately, many airlines in crisis situations respond to customers using sterile, scripted language. Many of these responses reflect poorly on the brand, and the customers perceive the tone as "a robot in a suit." These responses damaged the relationship between the airline and the customer.

Your social team isn't just managing public relations; they're shaping your brand's voice and its personality in real time.

Phone Calls and Hold Music: The Forgotten Frontier

Some companies treat phone calls as relics of the nineties. This is a dangerous position to take, especially in today's digital era. However, here is the truth. When a customer takes the time to call you in 2026, it is either because they are desperate or loyal. Either way, the stakes could not be higher.

If the customer is met with fourteen minutes of shrilling hold music, an eight-layer IVR menu ("press one if you'd like to scream into the air..."), or an uneducated representative reading from a policy manual, you have converted a potentially positive, loyalty-building touch-point into a break-up call.

Train your representatives to listen before solving, to personalize, and to resolve issues quickly. View each call as your last opportunity to save a customer's loyalty or lose it forever. Imagine if you strived for first call resolution or right-the-first-time on every call. Would the world be different? Absolutely.

I can't begin to share with you how many times I said to myself, well that was a wasted call!

The Words Between the Words

It's not only what you say. It is how you say it.

Compare "Unfortunately, that's our policy..." to human-speak: "I understand this is not the response you were hoping for, but let's see what we can do to assist." Amazon is an excellent example of human-speak.

Same facts. Completely different emotional impact. Let's examine some more examples.

I used to travel weekly for two years from Manchester to Brussels for work. There were two airlines at the time that flew that route. One had better times going out and one had better times returning. So, I got to experience them both. One time, coming home from Brussels, I went to check in and saw my flight was canceled. I approached the desk to ask to rebook another flight. I was met harshly and quickly: "No." There were no options, no way to get home, no flights leaving that evening."We'll let you know when we know." The thought of sleeping at Brussels airport or missing the family gathering the next day filled me with anger and severe sadness. That's when I shouted at the agent, "You are the worst airline!" And without delay, he shouted back at me that I was the worst customer. It would have been funny if he were not so serious. I retracted with my tail between my legs, looking for someone who could help me get home, even if that meant buying another ticket. Eventually I did, and arrived home at 1 a.m. the next day, but the memory of that airline and that agent has stayed with me for eighteen years. And to no surprise, that airline is no longer in existence—the first time into administration and the second time into full bankruptcy.

I visited a large grocery store chain on a Sunday morning to purchase basic staples (eggs and milk). At that time, both items were soldout. I asked a young employee for assistance, and she replied, "You should have come in earlier." I have remembered this for nearly fifteen years. Although the comment was somewhat amusing, it was not intended to be funny.

And because I have been obsessed with customer service and care since I was six, I could write a full book of examples. But instead, I share the most important insights.

Every interaction, regardless of size, contributes to the overall image of your company. Examples of touch points include clicking on a website link, not responding to an email, responding to a social media inquiry, and every sound of hold music and tone, every telephone call, every sentence spoken by your employees.

You are judged and evaluated by how much you care, not by your intentions. And your customers are paying attention to each touch-point, keeping score, and breaking up with you.

Part II–
YOUR CULTURE IS YOUR ORGANIZATION'S OPERATING SYSTEM

Your strategy, your tools, your technology—none of it holds without the foundation underneath. Part II is about that foundation. It is called culture, and it is the operating system of everything that follows.

CHAPTER 5:

CULTURE EATS EVERYTHING FOR BREAKFAST

You can write a new script for your team. You can write new policies or redesign them. You can remove and rip out tools, platforms, or apps and replace them with others. You can write playbooks upon playbooks. I have written many throughout the years. You can apply AI to a problem so many times until your CFO starts to smile again. But if your culture is toxic, dying, or doesn't exist, then all the other initiatives will be forgotten. Nothing else matters. I repeat, NOTHING ELSE MATTERS—no single initiative, investment, or transformation roadmap will make your organization better on its own.

There's an old saying in business: "Culture eats strategy for breakfast." I'd take it one step further: Culture eats everything for breakfast.

First, let's look at reality. Most businesses perform the same dance when their customer experience is poor: they blame the frontline, the tools, the economy, their employees; they complain, "our systems suck", and "customers these days." After that, they have a meeting. Then, a fresh, super fancy slide deck is released. Does this sound vaguely familiar? There may even be another meeting and another slide deck, and then a new project or initiative or tiger team is announced. Believe me, I have been there, done it, got the T-shirt several times. I have a collection of them.

And despite the effort and elegant PowerPoint decks, the experience doesn't really change. Because the issue was never the project or the energy behind the project, but instead the operating system or operating model.

Culture is not a banner of your values, or a poster of your ideals. Culture is the behavior employees show up with each day. It's how they live in the operating model system or office or contact center or boardroom. When employees are under pressure, when a customer is frustrated or upset, when systems or tools malfunction and calls in the queue blow up, or average handle time continue to be the only measures supervisors and team leads manage as a life-or-death metric, the culture of your organization is what shows up!

Culture needs to be what your company rewards. Your company's culture encompasses not only the rewards, but also what it tolerates and prioritizes in the approach.

What you reward as an organization is what will be delivered. For example, people will rush if you reward speed. If you criticize and punish those for thinking outside the box and going the extra mile to help the customer, they won't. If you encourage your employees to only follow processes and not allow them to use their judgement, customers will feel like they are talking to or working with a robot, or talking to a wall. Extensive rigidity is the cause of breakups each day. Celebrating and tracking risk avoidance never allows your employees to come up with solutions.

I know I sound a bit old here using this cliche, but if I had a penny for the number of times an agent has told me, I'm not allowed, I can't do that, or I'll get fired for this or for that, it would have funded a nice shopping spree and a fancy new handbag.

Therefore, culture consumes your strategy for breakfast, your newly developed initiatives for lunch, and your tech plan or roadmap for dinner. Your strategy deck is not what is experienced by your customer. They encounter your operating system or model by human interaction—or absence of one. And they do that through the way you accept or don't accept responsibility for when things go wrong; for your mistakes; whether you make things easy; if you are difficult to do business with, and if you care enough to do what you say and say what you will do and allow your staff to make the decisions on their own.

This is exactly the reason two businesses might use identical tools and produce radically different results. As a result of leadership fostering a culture of empathy, urgency, accountability, and ownership, I have witnessed organizations and companies with average to poor platforms or systems still provide exceptional service. I have also seen companies or organizations provide awful, annoyingly robotic experiences despite having top-tier tech stacks because of their fear-based, approval-driven customer-destructive mentality.

I assure you that a culture issue cannot be solved by automation. You can't automate your way out of anything. I know this first-hand. I tried it. In 2019, as I touted around the organization with the board of directors, I said, "Let's just automate it." I must have said that hundreds of times. I meant it, but what I didn't fully realize or appreciate at the time is that automation only elevates or scales where you already are. If you are broken culturally, it will elevate and scale that frustration, that internal pain, and carry that forward for you.

You need to finally acknowledge that culture is not being soft. I used to think it was fluff. But you must admit that it is anything but soft before you can delve further into people, training, and employee experience.

Culture isn't built in HR manuals or departments, laminated value statements in a booklet, employee surveys, or company handbooks. Culture isn't built in HR documents, quarterly surveys, or mission statements in the lobby. It's not fixed with a consultant or a rebranding or a new set of laminated values on the wall.

Culture is created and destroyed by the daily decisions leadership makes, often when no one is watching. How do leaders show up in crises? How do they respond to an angry customer or to a client? How do they treat their people on the frontline? The culture your leadership creates is vital. You can't "policy" your way into empathy. And you definitely can't "AI-chatbot" your way into genuine care.

Culture is not a frivolous thing. It is not a "nice to have." It is a business operating model or structure that is a necessity. Culture either fosters loyalty or subtly destroys it.

This chapter has provided a confrontation with reality, i.e., intention without culture is meaningless.

The question Chapter 6 will ask you is this: When your leaders show up under pressure, when a customer is difficult and a policy is easier than a judgment call, when the numbers are tight and the shortcuts are tempting—what does your culture actually do? That answer is your operating system. Right now, today, it is either working for you or against you.

CHAPTER 6:

WHAT LEADERS DO BECOMES CULTURE—FROM THE TOP TO THE FRONT-LINE

There is a dangerous misconception in most organizations that culture is "set at the top" and then somehow miraculously permeates the rest of the organization. This alignment of leadership takes place in a boardroom, is translated into PowerPoint, is summarized in a town hall, and ultimately reaches the front-line unaltered.

Culture doesn't work that way. Diluting takes place. Something is lost with each layer it moves through. Priorities get hazy. Bravery wanes. The context is lost. Only a small portion of leadership's objective is left when it reaches the front-lines. Ten percent, maybe twenty on a good day. Fear, pressure from middle management, antiquated regulations, and unwritten guidelines about what is "really safe" engulf the rest. This explains why a lot of businesses have a fantastic sound at the top but a bad feeling at the bottom. True culture is not hierarchical. It is both top-down and bottom-up at the same time.

Everything that doesn't belong is squeezed out when leadership expectations are reinforced both upward from the front-lines and downward from the boardroom. Poor conduct, thinking too slowly, defensive measures, and toxic rewards to name a few. The center is squashed when such alignment is absent. People experience burnout. Good people quit. Bureaucracy increases. Customers sense it. Culture exists in the middle. And it typically dies there.

Costco is among the best examples of culture living on the same page, wherever you are organizationally. Jim Sinegal, the former CEO, did not use slogans to create culture. He created it by being clear. Costco used what was effectively a merchandise doctrine for years, the Six Rights: "the right merchandise, in the right place, at the right time, in the right quantity, in the right condition, at the right price."

They also built the leadership framework most often attributed to Sinegal around five core principles (sometimes called "rules" or "priorities.") They are remarkably simple:

Costco's Five Leadership Principles

1. Obey the Law
2. Take Care of Our Members
3. Take Care of Our Employees
4. Respect Our Suppliers
5. Reward Our Shareholders

And Sinegal always added: "If we do those five things, our shareholders will be rewarded." Shareholders were the *result*, not the priority.

There were no buzzwords, corporate poetry, or strategic pillars, just values that you can run the company on. And it was intentional. No drama. No argument. And not to be reinterpreted. That served as the arbiter for all the company's decisions. It's not a spreadsheet. It's not a projection for the quarter. It's not a PowerPoint deck. Just simple guidelines. Most companies build it the other way. Costco reversed it intentionally.

Because Sinegal didn't just post these on a wall. He truly lived them. Especially when visiting stores and wearing his name badge: "Jim." He demonstrated listening first, protected wages and benefits, and refused the pressure of short-term Wall Street. The culture wasn't theoretical. It was modeled.

Whenever executives debated, it had to align with the five principles. That was leadership clarity and culture in action. One of my favorite facts about Sinegal's leadership is that he didn't manage from a nice corner office. Instead, he kept a folding table and a simple chair in a warehouse. He would always visit the stores, where he walked the floor, shook hands, and asked employees their opinions and truly listened. That is presence.

Why that was so important is that it sent a powerful message that people mattered and are more powerful than any corporate email or memo ever could be. Costco paid higher than market wages, invested in benefits where others were cutting them, and promoted from within. What became super clear was they trusted employees and protected employee rights.

Costco routinely touts the highest employee retention rates in retail. They demonstrate extensive product knowledge, high employee engagement, consistently strong service, and relentless member loyalty. This is all because of respect, not perks. Their famous return policy wasn't "generosity." It was an extension of trust. Employees weren't afraid to do the right thing because leadership had made it safe to do so. That is culture. It wasn't top-down speeches and bottom-up chaos. It was pressure from both sides.

Leadership set the values from the top. Front-line reality reinforced them from the bottom. Everything in the middle had to align—or get squeezed out. Bad behavior didn't survive. Bad policies didn't last. Bad incentives didn't stick. Because they didn't fit the system.

That's why Costco and Sam's Club look similar on paper but feel completely different in real life. Same category. Same warehouses. Same products. Different culture. Different leadership. Different results. Costco didn't win because they had better shelves. They won because they had better priorities. And those priorities were modeled every day by the person at the top.

This is the actual appearance of top-down and bottom-up alignment.

Let's get something straight: Your customer service is only as strong as the people leading your company and the alignment of all your employees. You can have the best product team, the most advanced technology, the slickest marketing, the best contact center, the best vendor, the best website, and the best BPO partners. But if your leadership doesn't prioritize people—your employees and your customers—all of it crumbles. And if your employees don't feel valued, respected, and trust the leadership, it doesn't work.

Leadership doesn't just set the goals of an organization. Leadership sets the tone, the permissions, the fears about what is safe and what is not. I used to always say, "Let's do the right thing." Leadership agrees with what is most important—doing the right thing for your customers and protecting the organization or company at all costs.

You certainly can't delegate culture to middle management while you focus on earnings. You can't delegate culture and outsource your values. You can't automate trust. You build it by how you show up each day.

When leaders protect people, people protect customers. When leaders trust employees, employees earn customer trust. And when leaders play the long game, customers stay. It is not soft, but good strategy.

The true masks come off when a crisis hits, when someone goes outside of policy, when the financials are super tight and cost-cutting has been identified. This is when leadership and culture show up. Leadership doesn't show up in all hands meetings or in Town Hall meetings when the quarterly slide says "Customers First," when behind the scenes, the organization is actually rewarding the complete opposite.

Leadership Drives Culture. Culture Drives Experience: The Real Test of Leadership

Forget the branding on the wall. Forget the values that were in a notebook. The real test of leadership and organizational culture is fully displayed when customers complain loudly, employees break the rules, and revenue is down. You can beat up your teams, or you can deliver motivational and encouraging sessions to get back on track.

CASE STUDY: TONY HSIEH AND ZAPPOS

When the late Tony Hsieh led Zappos, he didn't just talk about service, he embodied it. I was fortunate enough to meet and spend a few days with Hsieh and his team in 2010. And what I saw at Zappos was 100 percent created by his leadership, his beliefs, his ways of working. Everyone I saw, shadowed, and met were truly customer-obsessed, and not because they were told to be. They mirrored the fun and customer-obsessed leadership that Hsieh lived day to day.

He and his entire leadership team famously took customer calls. I remembered seeing the team's working area and honestly saying to myself, *That is the leadership*? I truly never saw anything like it. All open offices; it was like walking into a fun party, filled with things on the walls, and noise, and living their value statements. Leadership empowered employees to stay on the phone as long as necessary (record: over ten hours!). They rejected scripts. The leadership made customer obsession the company religion. And guess what? That culture didn't just "boost morale," it built a billion-dollar brand. They trusted people to *be* people, and in my meeting with HR, they shared that with me. I was in awe. Trusting your employees to be human? Really? Sitting there listening to the agent I was shadowing showed true connection with the customer who was ordering a pair of boots for an upcoming snowstorm. The connection with the customer was something I had never seen. Agents were encouraged to ask about the noises in the background, like, "Is that your dog?" or "What is the weather like in Montana today?"

This wasn't some PR stunt. It was daily leadership behavior. And the result wasn't just "happier employees." It was a billion-dollar business built on a reputation that customers would defend like family. That loyalty was their culture. This did not happen because of the slogans they discussed; their core values were built on key principles. And leadership modeled what mattered and so did the rest of the organization.

While that was many years ago, I often ask myself now if that leadership is different, is the culture the same? Are the values still modeled in the same way? Did they truly stand the test of time?

APPLE'S GENIUS BAR: CULTURE MADE PHYSICAL

Apple didn't invent customer support or tech support; they reinvented it by turning it into an experience. The Genius Bar wasn't just a table with smart people or a piece of nice furniture in the store. It was a cultural statement, a cultural move driven from the top. They made support not just a back-office function. They said, "Let us support you as part of our belief in our product. Let us help you even if nothing is wrong." Who does that today?

Steve Jobs understood that customers didn't want to feel stupid—they wanted to feel seen, helped, and in control. The result? The Genius Bar became a case study in customer trust and drove in-store traffic, loyalty, and revenue for years. Apple built a culture where tech support was elevated, not hidden. The Genius Bar became a place where expertise met empathy, and tech support felt safe asking questions they needed to enhance their user experience.

That decision, driven from the top, turned dreaded "IT support" into an experience customers lined up for. It drove loyalty, in-store traffic, and repeat sales. Culture created not just satisfaction, but lifetime value (a prediction of the total revenue a business can expect from a single customer account throughout their relationship with the company).

Whatever we think of Apple today, we cannot fault their product and the service and support that are available to us as customers. And when you enter their stores, you feel important and valued, and employees are happy to share what they know about their products. This is all part of their culture.

When you have toxic leadership, service becomes defensive.

Now contrast the Zappos story with a grow-at-any-cost culture. This culture tends to demonstrate a priority over the KPIs rather than the human element. It prioritizes image and popularity over accountability, speed and efficiency over humans. A global technology platform that connects riders with drivers during the early period was well known for how leadership sets the tone and shapes external trust. The culture at the top was aggressive, competitive, and tone-deaf. Customer complaints? Secondary and handled with defensiveness. Service teams were under-resourced. Leaders modeled growth at any cost. It was inevitable that a fallout would happen, particularly as they earned a reputation for toxicity, which took a long time to repair.

Even after leadership changes, organizational resets and trust can take time to build, and service reputation can remain fragile—all because customers don't usually forget how you made them feel, ESPECIALLY when they had a problem and you met that problem with remiss.

Customers may even argue that the brand's service reputation still has ground to cover.

If leadership culture screams "revenue first, people last," that attitude spreads everywhere into how customers are treated, how employees behave, and how your brand is perceived. Remember that!

Culture and Leadership are Contagious

Employees don't copy posters. They copy leaders. When a leader stands up at an employee kick-off or a conference and says what he or she truly believes, you can bet your bottom dollar that employees will talk about that for the next year. That sets the tone, the pace, the practice.

If executives and leaders only talk about cost control and quarterly earnings and ignore service issues, employees learn quickly that *cost and revenue matter, and customers are optional.* If leaders celebrate empathy, urgency, and ownership, employees will mirror that too.

Leaders don't just set KPIs. They set the emotional thermostat for the whole organization.

Leadership is defined not only by what one says, but by what one allows to happen.

I previously collaborated with a leader who often asserted, "We are merely selling hotel rooms." I countered, "No, we are marketing experiences." For certain guests, this constitutes a singular, unparalleled journey. The disparity between leadership perceptions and customer sentiments is the fracture where culture deteriorates, with frontline staff as the initial witnesses.

If representatives are penalized for taking an additional minute to resolve an issue, they will cease problem-solving activities. If leaders dismiss grievances, the organization adopts a similar attitude. If "policy compliance" is perceived as safer than ethical action, your culture will subtly condition individuals to prioritize safety over service.

Tone-Setting Example: Airbnb During COVID

When COVID shut the world down, Airbnb refunded millions in canceled reservations through its extenuating circumstances policy. They didn't have to, as eligibility depended on when the booking was made. It greatly cost them. Hosts were furious, but guests remained loyal.

That decision wasn't a financial spreadsheet maneuver. It was a cultural move. It said: *We put trust before short-term profit.* And Airbnb's brand survived a near-death experience.

Walk the Walk and Talk the Talk

If you're a leader reading this and haven't walked the floor in months, haven't answered a customer complaint personally, or haven't shadowed your contact center, start there. Otherwise, you are managing culture from afar. I assert that isolation fosters delusion.

Experience your business the way your customer does. Try buying your own product. Then try returning it. Try to find help on your website and call your own customer service number regularly. Navigate your own IVR like a customer who is already frustrated.

If any of those process experiences are clunky, cold, or confusing, fix them. More importantly, show your teams that you care enough to do it yourself. It always amazes me how many businesses force you to call because the online process didn't work as it should. Then, because businesses put all their eggs in another channel, the incident is even worse for a customer after an already failed digital experience. By owning your own customer experience path, regardless of channel, it will reset and set your culture thermostat, and your teams will start to match your behaviors. Go for it.

Empowerment Starts at the Top

Your frontline team cannot deliver great customer service if they're not trusted to use judgment or they need a manager's blessing for everything or if breaking policy is deemed as career suicide. Empowerment doesn't mean chaos; it means trust with clearly defined guardrails.

Employees should feel like they're hired to help people, not just execute tickets. That shift begins with leaders giving permission to solve real problems without various levels of override or approval.

A Modern Proof Point: iQor's Cultural Refresh Case Study—Building a Connected Culture That Powers Experience

iQor Inc. was at a crossroads when a new CEO took over. Fresh equity partners had come on board, and the company was growing rapidly. Branding had recently undergone a refresh, and the company had developed innovative AI technology to revolutionize the global customer experience.

From a purely financial perspective, all the pieces were falling into place. However, leadership recognized that something was missing. Although technology, funding, and branding can create a basis for growth, without a solid cultural foundation, such growth will be unstable. People may accomplish their tasks, but they won't be working together. Teams may perform well, but they won't have confidence in their actions. Ultimately, clients and customers see the difference.

This awareness served as the impetus for one of the largest and most impactful undertakings in iQor's history, the total resetting of the company's Mission, Vision, and Values (MVV). I had the honor of leading this effort with the company's chief culture officer during my tenure as an executive. We recognized this would not be an endeavor to update a message or rebrand the company, but to define how the company viewed itself, and how it would function, provide service, and grow.

The executive team recognized a simple fact. Culture is not an HR initiative, nor is it a statement in an employee handbook. Furthermore, it is NOT a slide presentation. Rather, it is the "air" employees breathe every single day. It is the determining factor of whether employees will show up to work to do a task or have an impact. To achieve the goal of establishing iQor as a leader in both innovation and humanity, the company's culture had to be as deliberate and structured as the company's strategy. Therefore, it had to ensure the company's culture was as easy to recall as the company's strategy, included all roles, represented the desired impact the company wished to have, and was centered upon the needs of employees, clients, and customers.

This was not simply about putting words on a wall. It was about unifying the approximately 50,000 employees in various locations, roles, and departments behind a unified identity and purpose.

Timing was also critical. The organization had experienced a tremendous amount of change in recent years. More than half of the employees had less than two years of tenure, while many other employees had worked for the company for more than ten years. When leadership asked employees from various departments throughout the organization what the company stood for, more than 400 of them could not clearly describe the current mission or vision. While most employees could identify some of the values that existed, due to the number of values and the lack of clarity, no employee could consistently identify them, let alone demonstrate them in their daily behavior.

That disconnect was the sign; not of failure, but of incredible opportunity.

Instead of viewing the MVV refresh as a top-down proclamation, leadership chose a much more challenging path. First, leadership aligned themselves at the executive level on what the future of iQor would require in terms of a new chapter. Then, leadership invited the entire organization to participate in the process. Regional leaders, agents, and supervisors around the globe were engaged early in the process, at all levels, and in all geographic areas, so that the efforts would not appear to be originating from corporate headquarters.

Employees at all levels, including front-line agents, supervisors, and managers, participated in the creation of the new mission, vision, and values, so the values would represent the actual experiences of the employees who provided services every day. All employees, including those from operations, HR, technology, sales, client services, and support teams, as well as the executive leadership team, collaborated during the process so the language used would be inclusive and able to transcend departmental silos, rather than reinforce them.

Listening sessions, working groups, and actual dialogue occurred as a result. Employees did not merely receive the company's new culture; they assisted in creating, or as I like to say, co-authoring, and developing it.

The outcome was not a collection of aspirational phrases. It was an operating paradigm. The paradigm was clear enough to remember, real enough to utilize, and grounded enough that employees could reference the model when making difficult decisions and understand how the company intended for them to act.

However, leadership did not merely develop the words to describe the culture.

Leadership established the structural components necessary to keep the culture alive and sustainable. The Connected Culture Committee became a permanent committee that maintained direct access to the leadership team and the authority to influence the decisions of the company. It created a continuous feedback loop between the frontline and the executive team, and it ensured that the culture that was verbally communicated was consistent with the culture that employees were experiencing.

During the same period, programs such as WIN (Women's iQor Network) transitioned from concept to reality. These programs were not symbolic gestures, but examples of the new culture's commitment to inclusion, growth, and respect. Programs such as WIN provided a platform to elevate employees' voices across the organization. It created developmental paths for emerging leaders. It clarified that empowerment was not a secondary initiative or a slide presentation. Empowerment was real and it was the culture.

As these changes began to occur, the results were undeniable. Executives and employees began to communicate using the same vocabulary. Employee engagement increased as employees connected with the purpose of the company rather than solely performing tasks. Clients perceived the shift in tone, ownership, and the humanity of the experience. Quietly, culture has become one of iQor's greatest competitive advantages.

Perhaps most importantly, the company learned what great leaders already recognize: Culture is not an abstract concept. Customers and clients can feel it. They hear it in the voice of the agent. They experience it in the fairness of the process or in the way the company conducts its business. They sense it in whether the company operates within its values when things go wrong.

What occurred at iQor provides a template for any company that wants to take culture seriously. Change must begin with leadership who truly believes the value and the importance of it. Every level of the organization must be involved so the values reflect the reality of the organization. The work must be reinforced by ongoing structural elements, not one-time rollout events. Inclusion must be demonstrated and lived by the company. Lastly, culture must connect directly to the customer experience, because if customers and clients cannot feel the culture of the organization, neither can employees, and it does not exist.

Most companies speak about culture. Fewer still operationalize it.

iQor did. And I was fortunate enough to truly help lead in driving this effort. This illustrates the fundamental lesson of this chapter: Culture is not an accessory to business. It is business. It is an invisible operating system that enables customer experience, employee engagement, and long-term growth.

Technologies will evolve. Business strategies will change. Markets will fluctuate. However, the culture that leaders establish—or don't—will determine whether the company succeeds or gradually deteriorates. Your leadership and your culture are your operating system or model. What you prioritize as a leader will ultimately define your company's reality. Strong service requires visible, committed, and aligned leadership. Culture is not a benefit. It is your business model.

CHAPTER 7:

HIRE FOR HEART AND EXPERIENCE, AND TRAIN FOR SKILL

You can certainly teach people how to log a ticket, use a CRM, or perform refund policies, escalation procedures, or technical troubleshooting steps. You may even be able to script the first thirty days of their employment. But you can't easily teach someone how to really care.

If you look back at any amazing brand that once represented something but then slowly lost its way, you will most often see the same tale of woe. The leader who once lived and breathed the company's culture was replaced by someone who could manage it, measure it, or monetize it, but did not embody it. The values did not disappear overnight. They faded when they stopped being modeled. And when leadership stops living the culture, the culture stops living within the organization.

At its core, customer service is a people business. It always has been, and I truly believe it always will be. While systems, strategies, and technology matter, they are only as effective as the people using them. Each interaction, each smile, each pause before a reply, and each empathetic tone determines whether a customer leaves feeling elated or disenchanted. Will they become loyal or switch to your competitor's offering? Will each interaction lead to a positive experience or a negative one? Those answers depend on the representatives of your brand. And the best brands know this. They hire people who care and train them to be phenomenal.

Your people, from the frontline to the C-Suite, are your brand. If they do not care about people, everything else you build will be irrelevant, I can assure you of that. Regardless of how sleek the automation or advanced the AI that accompanies your people, if your people do not care about people, the rest will not matter.

Hire for Heart and Experience. Train for Everything Else.

Attitude trumps aptitude every time when it comes to hiring for frontline positions. While you can teach someone to follow a playbook, it is much more difficult to teach the type of heart that instinctually wants to help people and genuinely cares about doing so. How many times have you said, "If only they thought for themselves?" Again, if I had a nickel... another new handbag for me!

The ideal candidates for frontline roles typically exhibit certain fundamental qualities. These attributes were among the qualities I sought when recruiting agents. Ideally, they possess an innate interest in understanding the mechanics behind operations, the emotions of customers, and ways to improve customer experience. They continuously strive to create improvements for both the organization and the customer. They demonstrate extraordinary listening abilities and exceptional listening skills and take time to formulate thoughtful responses. They remain calm during periods of difficulty, especially when customers are experiencing stress. They also exhibit resilience when working in high-stress environments. They tend to be more inclined to show empathy than self-importance. Egoism in any profession is hazardous, regardless of what level. They also possess emotional intelligence and recognize tone, timing, and the value of non-verbal gestures.

By hiring for heart, you are not simply filling a role. You are investing in a person who views assisting customers as an honor, rather than an obligation or a duty.

Below are two of my favorite organizations that exemplify this philosophy. When you walk into their stores or ride on their aircraft, you can undoubtedly sense that their employees are encouraged to be themselves.

TRADER JOE'S: GOOD PRODUCT, GREAT PEOPLE

Trader Joe's has created a fanatical customer following, which has garnered national attention without engaging in massive advertising efforts. Instead, they've cultivated a unique emotional shopping experience for their customers. The company deliberately selects employees for their personality, rather than prior retail experience. Why? The employees at Trader Joe's exude a level of warmth, humanity, and presence that you cannot manufacture. Associates are hired for their friendliness, humor, and genuine curiosity. They're encouraged to be themselves, engage customers in conversation about products, recommend their favorites, and brighten someone's day. Customers can tell when a store allows employees to be themselves and let their personalities shine.

Compare that to many other retailers or call centers that tell employees to leave their personalities outside the door. Trader Joe's takes the opposite approach. They celebrate individuality. As a result? A highly loyal customer base developed as much by their products as by how customers feel when they shop there.

SOUTHWEST AIRLINES: FULL OF LOVE AND FUN, BUT ALSO VERY GOOD AT THEIR JOB

Southwest Airlines follows a similar philosophy. They recruit for attitude, humor, adaptability, and human connection. In their hiring process, they prioritize friendliness and a human element. Flight attendants are famous for bringing personality to their announcements, handling angry customers with empathy, creating lighthearted moments throughout the travel experience, making impromptu announcements, and calming irate or frustrated customers.

Unlike the rigidly scripted announcements of competitor airlines that create robotic interactions, Southwest encourages their employees to prioritize real human connections that foster customer loyalty and retention and creates a brand identity centered around warmth and concern. They need employees who can think quickly, handle difficult passengers with kindness, and add personality to the entire travel experience.

I can honestly say that every time I have flown with Southwest, a flight attendant has made me smile through their announcements and interactions as they move up and down the aisle. It shows that they truly enjoy their job and the people with which they collaborate and provide services for. It gives the impression that they love what they do and the people they do it with and for.

Experience Is a Must for Leadership

Here is where the philosophy shifts. We also need to stop fooling ourselves. Let's not confuse front-line roles with leadership requirements.

Experience is clearly necessary for the various levels of leadership within an organization. The philosophy is different regarding experience versus other factors such as talent, charisma, and creativity. While you can hire a customer service representative, sales professional, or contact center agent with little or no experience, especially if they have a good attitude and a desire to learn, hiring a director, vice president, or executive with little or no relevant experience is completely unacceptable.

Knowledge in customer care, operations, and service delivery is paramount at the senior leadership levels of customer strategy (CX/CC/CE). Your senior leaders cannot simply be nice or inspiring; they must be builders and system architects. They must be able to align cross-functional teams, take customer feedback and turn it into actionable strategies, and be a visionary of structure and strategy and a champion of scalability.

At this level, senior leaders must be able to create scalable support frameworks that produce measurable results. They must have the experience to select and optimize the correct technology ecosystems for their organizations, without over-complicating them. They must be able to correlate customer metrics to business performance and inspire their teams through both empathy and excellence. I've worked for leaders who didn't even know the basic CX/CC/CE metric definitions. Their own teams had to explain that to them. You can imagine how that worked out, right?

Empathy remains critically important; however, it must be combined with the appropriate experience and operational knowledge. I am continually shocked to see how many major corporations hire executive-level professionals with virtually no relevant experience. In addition, many of these same executive-level professionals rarely attend industry conferences and therefore do not understand the complexities associated with developing a comprehensive strategy that drives service, revenue, sales, and workforce management.

I have witnessed far too many organizations hire or promote executive-level professionals based on their charisma or intentions, not their competencies. As a direct result, the organizations suffer from chaotic operations, countless amounts of spin, and delayed implementation of technology or automation, or frankly anything that would advance the business. Charismatic and intentional people should not operate businesses or manage operating models.

Otherwise, I have seen, the entire operation stalls. This includes delayed selection of the correct technology stack, stalled implementation of AI and automation, and ultimately, the inability to bring together people who are knowledgeable in products and technologies and ensure all cross-functional departments are aligned. I have also seen poor collaboration between cross-functional teams. Customers become angry due to a lack of proper preparation, empowerment, and support for their employees.

This became glaringly apparent while I was employed at a massive organization that had a significant focus on both revenue and customer experience. Despite spending millions of dollars on consultants and deploying limited technology, the organization continues to struggle with executing and implementation. Not surprisingly, those with the greatest amount of experience eventually leave the organization because it was a futile effort to work for there.

Lastly, if the executive-level professional lacks in industry credibility, the organization will not be able to establish the proper culture to follow. As stated previously, the middle layer is vital to setting up the proper culture for your service, and if it isn't established at the top, the middle layer will be unable to guide the culture properly for the rest of the organization.

The Middle Layer: Where Service Cultures Live or Die

Having extensive experience working for large- to medium-sized organizations, I truly believe that most organizations greatly underestimate the importance of the middle layer. Most often, the biggest breakdown occurs at this level: team leads, floor supervisors, and regional managers. These are the people who translate the executive leader's strategy into everyday action. The middle layer acts as the link between the strategy and the execution. If the middle layer is disenchanted, disenfranchised, or poorly trained, your customer experience will suffer and your culture will perish.

Your frontline will also fail regardless of how great the vision is from the top. Trust me. I learned the hard way. Regardless of how long you search for the ideal leader at the top, you may still see performance falter because the everyday coaching layer is the true engine of the organization. That engine will never be able to provide the type of coaching to help your frontline reach a state of excellence if it is not designed to coach people. Someone must be passionate about people to be successful in a service-oriented role.

Middle managers who are exceptional must be aware of how they present themselves to their teams. They must be coaches and conflict solvers prior to being task-focused. They must be able to identify and repair processes, not just enforce rules. They must also be able to solve problems pragmatically so they can find a balance between adhering to policies and providing the flexibility that is required to maintain a loyal customer base. I will discuss this further in the book.

Train Relentlessly

Training also needs to be mature. Training is not a two-day PowerPoint extravaganza followed by "Good luck." Training develops muscle memory. It is scenario-based training. It is emotional intelligence. It is having the confidence to perform under pressure. It is instructing employees how to say no in a manner that still demonstrates respect. It is teaching employees how to recover when things go wrong. Zappos did this better than anyone I have ever observed, and that was before simulation and AI. Now we have even better tools to help with this.

If your training is not preparing employees for real-life situations, then it is not training. It is paperwork. The quiet truth is that customers do not lose their temper with your employees because they are bad; they lose their temper with employees because they were never provided the tools, the empowerment, and the support to succeed.

Zappos Onboarding Engineering Culture Before Proficiency

From my visits to Zappos in the past and conversations with industry peers, Zappos still uses a deliberate, culture-first approach to onboard their employees—it's famous for it.

New hires to Zappos view culture not as training time or as a poster on the wall, but as the operating system of the organization.

Every new employee, regardless of position, participates in a multi-week onboarding and training program focused on the company's mission, history, and ten core principles. The goal is to educate employees on Zappos's definition of success: to deliver WOW to customers through customer service, teamwork, and consistent behavior. Zappos believes the behaviors of how the employees interact with each other come prior to the behaviors of how they complete their job functions.

Culture is not a solitary lesson; it is perpetually reinforced through continuous education, leadership examples, and a shared vocabulary across the organization.

One thing that makes Zappos unique is their emphasis on protecting the cultural alignment of their employees. The company received attention for offering incentives to new hires to leave immediately upon completing their onboarding period if they decided the culture was not for them. While specifics surrounding this tactic are not always readily available on their career page, the premise is clear: Zappos wants to invest in people who will authentically adhere to the company culture and not simply employees who are driven by the job itself.

Culture is truly valued at Zappos. It is taught, selected, evaluated, and constantly reinforced, long before job titles, performance metrics, or production expectations are discussed.

The proof of that culture doesn't live in the onboarding manual but instead shows up in moments no policy could predict.

During one of my visits, I witnessed the kind of interaction this training produces. A woman called to return shoes she had purchased for her deceased mother. Zappos processed the return without hesitation—and then went further. They sent the woman a check for the full cost of the merchandise, delivered flowers to her door, and offered their condolences. No manager approval. No escalation. No special protocol.

This wasn't a budget decision. It was a human one—made by a frontline employee who had been hired for heart and given the permission to use it. That's what culture-first onboarding produces: people who know what the company stands for, believe it, and act on it when no one is watching. Every representative at Zappos was empowered to respond exactly this way—because that's who they hired, and that's what they trained for.

Contrast that with an airline representative who, when a customer complained about lost luggage on a transcontinental flight, suggested she should have packed a toothbrush in her carry-on. That incident became a viral sensation—not because of the lost luggage, but because of the callousness of the response. The representative didn't fail a service standard. She revealed a hiring and culture failure. No one developed with genuine empathy and real empowerment says that to a frustrated customer.

Intentionally Developed Teams (Read this Twice)

Excellent customer experiences do not occur from the best-qualified or most-credentialed teams, but from the most intentionally developed ones. These consciously constructed teams are formed by hiring and selecting employees who have an empathetic and curious nature and a predisposition to care about their customers and the work they do.

Training and enabling employees with the skills, confidence, and resources to perform at an optimum level enhances these types of teams. And they continue to exist through senior leadership experience and the ability to design systems, manage and scale service, and transform empathy into actions.

The Zappos employee who sent flowers to a grieving customer didn't check a policy manual. She didn't ask for approval. She acted because she had been hired for heart and given the permission to use it. That is a hiring and culture decision made long before she ever picked up the phone.

Before you move to Chapter 8, ask yourself one question: Are the people you hired for your frontline roles there because they genuinely care about other people, or because they were available and the position needed to be filled? Your answer to that question is your service culture. Everything else is just the container.

CHAPTER 8:

STOP IGNORING YOUR EMPLOYEES AND YOUR PRODUCT AND DESIGN CHOICES

Here's the painful truth: If your employees are unhappy, disengaged, and burnt out, your customers will feel it long before you do. If your leadership team lacks trust in one another or believes there is no vision or strategy, your customers will see that. Your people are your first customers. Every frontline experience is a reflection of your internal culture. You cannot create amazing service on the outside when you treat your employees poorly on the inside. Customer experience never fails by accident; it fails by design. Not product design, not system design, but culture design.

If your service team is overworked, underpaid, undervalued, and unheard, then you're not just bleeding talent—you will be bleeding customer loyalty.

Your people must be your first customers. Your product and service design are the tools you allow them to utilize, and your leadership choices determine whether they are empowered, engaged, or just trapped.

That is why you cannot build great customer experiences on a broken employee experience. Equally, you cannot succeed with your employees when your systems are broken.

For years, I remembered that grocery store checkout moment as a major and epic customer failure. The cashier who challenged my mother's integrity—the tone, the suspicion, and the feeling that we did not matter—didn't care. But after decades of leading organizations, building service operations, and growing talent in cultures that allowed me to be successful and in others that didn't, I faced another epiphany.

Perhaps the cashier was not the problem. Perhaps she was the product of a larger culture issue at the grocery store, and she became the designer of that company's operating system. There, I said it out loud. I can now go back to that store that I refused to ever shop in for all those years.

She could have been the product of a culture that prioritized loss prevention over human dignity or care. The outcome of policies punishing adaptability. The outcome or product of systems that valued compliance over connection.

Writing a book can be therapeutic... I honestly never thought about that until now.

She may have previously faced punishment for overlooking an issue or letting something slide. Maybe her performance metrics rewarded stringent enforcement above resolution, and the environment in which she operated conditioned her to distrust first and care second, or even last.

That six-year-old me felt the profound effects of that leadership decision. This can be exactly what happens when leaders design or allow cultures, products, and policies to protect the business at the expense of human experience. Employees embody these shortcomings and become the visible face of invisible failures.

Every bad customer experience is the outcome of someone executing a system you approved of. When your service people become overburdened, underpaid, under-appreciated, and ignored, you won't be just losing talent, you'll be losing customers. Customer experience makes a mockery of your lack of patience, your chilly tone, your three-minute mechanical dialogues, your third and fourth transfers, and your little invisible box of resigned "that's the way we do it here."

Your product and service design make it easy for your people to care, or they make it impossible to care. And the data show this.

The Employee–Customer Connection

How you treat your employees is how your employees will choose to treat your customers. Fancy secret. I know. The Costco case study illustrates this, but really, it's a beautiful truth that so few companies know.

Gallup research shows that organizations with highly engaged employees outperform financial competitors in their industry. What does that mean? It means engaged employees take ownership of customer problems, they imbue their delivery with warmth, ease, and consistency, and they actually stay longer, retaining institutional knowledge and service continuity.

When employees feel like you don't see them, hear them, or value them, customers see it in shorter patience from agents, cold robotic tones, excessive hand-offs interactions, escalations, and of course, limited or no empathy.

Costco vs. Big Box Competitors

As we saw in Chapter 6, Costco's approach to employees produces measurable results—low turnover, deep product knowledge, and people who genuinely care. That philosophy shows up in customer interactions too.

I remember bringing a very expensive product back to Costco when a better one became available at a comparable price. While this was not within two weeks of purchase, there was never a no on my return. I was asked why I was returning the product, but I was never argued with. And I was treated well, like my request was reasonable. You might think this is a one-off or that I got lucky with that employee. But it's part of their customer promise and their 100 percent satisfaction guarantee. My receipt was also not needed because it could be tracked through my membership.

Now contrast that with many big-box competitors who churn through employees and treat them as disposable and design operations around cost control alone. This revolving door means customers encounter new, untrained employees every visit. Service becomes inconsistent. Frustration builds. Loyalty evaporates. Leadership blames "the job market," while ignoring the culture and the systems they built.

I'm exhausted just writing that.

I can guarantee you that most places do not have an HR issue but instead have leadership problems designed to call this reaction. When employees start saying, "We've been flagging this problem for months," "Leadership only listens when a customer complains," or "We find out about changes when the customers do," you no longer have a customer service issue, you have an employee experience failure. Your frontline is your early-warning system. Ignore them, and the first time you hear about the problem will be when it goes viral, costs you revenue, or shows up in attrition.

Too many organizations forget that frontline employees are your early warning system for customer frustration. If you're not harvesting those insights and have good AI data on your calls and VOC, I almost guarantee you are missing what's really going on in your company. The irony of all this is that these AI analytics are available now and are not expensive to use and could be gathering this data so easily.

I once worked for a very large utility company overseas. My team and I continued to get complaints about our ineffective and difficult customer account set-up process. No one listened. So, we came up with a "genius" idea to show the board of directors. We built a Customer Experience House in our unused ground floor and took the board through it. After they entered our home, we asked them to take off their shoes and then went through the customer experience of setting up a new account. They were horrified at how painful it was to sign up as a new customer. They had been clueless. Our efforts to teach them worked, and they listened. They supported us in making process and policy changes, which made it so much easier for customer account set-up, eventually supported by new systems and new automation. If those customers could have gone elsewhere, they would have. But because it was a utility service, they had no choice. If your customers do have a choice, though, you won't have the time to prove dissatisfaction. By then, your customers will have already left.

Employees as Brand Ambassadors

Your employees are not just executing transactions, they are living, breathing manifestations of your brand. When they feel that you are trusting and respectful and interested in them enough to help them succeed, they positively recommend your services or business, not because it's a phrase in some pitch deck, but because they genuinely believe in what they sell, engage, and advocate, and will go the extra mile without being asked.

When they feel neglected, overworked, or undervalued, the desire to care evaporates, and your customers feel it too. This also applies to leadership and management.

The Four Seasons is one of my favorite hotel chains because they don't just train staff to treat guests exceptionally. They empower managers to treat employees with the same level of care. Employee feedback is actively solicited and acted upon. The result? Turnover rates are far below the hospitality industry average, and service feels genuine because it *is* genuine.

Listen Like You Mean It

Many companies have "employee feedback" programs in name only. Annual surveys are useless if they're treated as a formality.

Effective listening necessitates continuous feedback mechanisms, hierarchical dialogues, and observable responses to be visible when employees speak up. When you feel heard, people listen. When your people observe change, your trust factor increases. When change doesn't happen after you speak, you stop speaking. When your team stops communicating, you silently start to lose your culture.

The number one unused retention strategy in a company? Recognition! Not plaques or trophies. Not some slogan emblazoned across a break room wall. Genuine recognition from superiors for enthusiasm, ingenuity, and problem-solving. But there are not enough ping pong tables, complimentary sandwiches, or casual Fridays that will compensate for poor pay, bad benefits, and inconvenient scheduling, for bringing people back to the office, or to remedy a lack of growth opportunities. Who is going to take care of customers when they are worried about making it to next week's pay date?

Middle management serves as either the bridge or the impediment within this framework. The customer experience is predominantly influenced by direct supervisors rather than by policy documents. If managers are disengaged or lack support, they cannot effectively advocate for their staff, resulting in the departure of top talent.

Disregarding your employees adversely affects morale and is directly detrimental to your customer experience. Investing in your employees' well-being, development, input, and leadership is palpable to consumers in every engagement.

Content employees foster customer loyalty. Dissatisfied employees generate turnover, disengagement, and negative Glassdoor evaluations. The accountability for all of it starts exactly where it always has—at the top.

PART III–

WHERE INTENTION MEETS EXECUTION

This is the part of the book where we discuss how good intentions become real experiences—and why, so often, they don't. It's not that people don't care. It's that caring has never been operationalized.

CHAPTER 9:

LET'S GET TACTICAL

One of the best lessons in execution doesn't come from a business book. It comes from one of my all-time favorite movies, *My Cousin Vinny*. Vinny Gambini initially appears to be a complete wreck. He is loud, unskilled, ill-prepared, and unappreciated. He is self-assured and well-meaning, yet he lacks obvious credibility. He doesn't seem like a candidate to win on paper. Nevertheless, he does. Not because he speaks more effectively, but because he listens.

He examines the data. He challenges presumptions. He takes tire-track measurements. He experiments with timelines. The narrative is inconsistent. The angles are ineffective. The facts don't add up. The case begins to leak. Suddenly, something that seemed solid is no longer viable. Then the pivotal moment occurs.

Vinny calls Mona Lisa Vito to the stand, an attempt to demonstrate that the prosecution's whole case is predicated on an erroneous notion. He asks, glancing at her: "Ms. Vito, could you please respond to this question? Does the defense's case hold water?" She doesn't think twice. "No! It doesn't. The defense is wrong!" Everything falls apart in that moment.

Not because the opposition was dishonest, lazy, or careless. Not because they didn't make an effort. Their execution was not in line with reality. They had a compelling story. It sounded good. Their logic seemed reasonable, and they had good intentions. But when tested, it didn't hold water.

This is exactly what happens every day in business. The courtroom serves as an ideal metaphor for the corporate world. Every business has a "defense" when it enters the market. "Customers are important to us." "Our customers come first." "Experience is important to us." "Service is our top priority." It sounds nice. On slides, it looks good. In strategy decks, it reads nicely. That's the goal. The customer then arrives. And they begin testing it, just like Vinny.

I had to call three times, but why? Why was there no follow-up? Why do your agent and website say different things? Why didn't you fulfill your promise? Gradually, your argument begins to disintegrate.

Because purpose is ineffective. It is executed. Your promises are at odds with your policies. Your service is compromised by your systems. Your employees are frustrated by your workflows. Accountability is broken by your handoffs. Customers therefore look for weaknesses in your "defense." And faith vanishes as soon as they notice the holes. Confidence without alignment is meaningless, just like in *My Cousin Vinny*.

Credentials are meaningless if they are inconsistent. If the facts don't support a decent story, it falls apart. Outstanding businesses are aware of this. They don't use catchphrases. They construct evidence. They don't just promote compassion; instead, they operationalize it. They make no claims to loyalty. They deserve it. Every encounter is an experiment. Every grievance is subject to cross-examination. And the same query is posed by every customer: Does the story of this company really hold up?

Are your promises fulfilled by your service? Do your systems uphold your moral principles? Are your employees equipped to deliver on your advertising? If not, the decision is straightforward. The defense is incorrect. And there is no appeal in business once clients arrive to that conclusion. They simply leave you!

We also learned that caring is not encapsulated in a mission or vision statement, but lives in your operational framework. It resides within the mechanics of service, the systems, the policies and procedures, the workflows and the daily decisions that dictate whether your company is easily accessible when doing business transactions or makes it difficult for your customers.

I saw this collapse in real time with one of the largest specialty apparel companies in the U.S. I have been extremely loyal to this portfolio of brands. So much so that my lifetime of spending could have bought me a nice Porsche. This group is in the middle of a massive transformation to drive efficiency. They are putting in new platforms, new technology, and what feels like a new operating model—overnight.

In practice, it has been chaos, and I complained several times as an avid fan and customer. They are doing about ten things at the same time to drive hard this transformation. For example, they decided their customer support agents would support all their brands and collection of lines of clothing with different products, new systems, and different rules and customer promises. They layered in various outsource partners across different geographies, they changed their shipping policies, and rolled out many new tools, trying to deliver this all in a short amount of time.

I felt their service getting worse, and when I shopped in-store, I noticed the store associates who needed to contact customer service felt it too. They were angry and frustrated like me. They couldn't get answers and were passed around from department to department, agent to agent.

From what I understood, agents were drowning—trying to master new products and services while processes changed regularly. Their loyal outsource partners were not trained well, and certain permissions were unavailable, which forced calls to hit escalation at times. After spending nearly $2,000 in a store one weekend, then waiting for certain items of clothing to be shipped, and calling customer care more than twenty times over just a few days, I lost my patience. Enough was enough.

Despite valuing communication, I still got stuck in the middle of their transformation nightmare, and I felt just like their own employees and associates—overlooked.

This company had the right intentions, but there was no logical progression of activities; they didn't balance out what was needed and/or safeguard customer experience.

I wrote to who I thought was the CEO, not as an angry buyer but as a ten-year, excessively loyal customer whose substantial value and trust in the brand had been softly demolished. The expedited shipping I paid for no longer worked in the store or online. My top-tier level no longer mattered. I tried to reach out multiple times, but could not get through to anyone in the U.S. escalation management team. I was told I'd receive a call back several times—I didn't—and when I asked to be transferred to a supervisor, I was told I couldn't be. Almost everything I loved about this brand—my experiences there, the beautiful clothes and accessories I had purchased—was changed. All at once, over a period of about six months. I had no path to resolution until I finally got connected to the senior vice president of customer care—by about the twentieth call.

Being in the industry, I understand how hard it is to do multiple transformations at once while keeping the business going. It's like trying to change the tire on a car while driving 100 mph down a highway. That's what it feels like each time the organization doesn't put a plan in place.

It was clear the SVP understood the mission, the plan, and the goal. It was clear he cared about what they were doing and understood the grave impact this was having on customers and employees. What totally stood out to me was that he was a product of what his organization was delivering, and he was struggling to put it right.

He said some things that still haunt me today. "I wish we could do that." "I wish we could fix that." "I wish we could support you in that way." He cared. I could hear it. But his system didn't.

What I learned next explained everything: They were "testing" new systems, combining brand support, changing shipping rules, and reworking operations all at the same time. Been there, done that. But at what expense were they transforming? Transformation is difficult, and efficiency is critical, but again, at what cost to your most loyal customers?

I write this next line as the platinum rule of execution. Execution is not about how many things you change at once. It is about whether the customer and the employee can survive the change and whether the change brings about the right outcome for them. Does it make it easier for the customer and better for the employee? Those are the only questions that should be asked when measuring its success!

This is what happens when companies confuse movement with progress. When they chase efficiency without steadying experience. When they stack change on top of change without redesigning the operating engine underneath.

Getting tactical means slowing down enough to ask the questions leaders avoid.

Does this make it easier for the customer? Does this make it easier for the employee? Does this reduce customer and employee friction or just move it somewhere else?

If the answer is no, you are not transforming. You are destabilizing and destroying your brand, your reputation, and your customers leave. End of Story. Full Stop. Period!

So here it is in a nutshell. Executives and leaders often confuse strategy and execution. These beautiful PowerPoint slides never survive in the real world. They assign this to a department rather than integrating it into business structure. They undervalue the difficulty of resolving years of process, technology, and ingrained cultural practices that subtly sabotage customers daily.

Here is what I've seen in thirty years of corporate experience, not everywhere but many places: new mottos, updated organizational charts, very pretty dashboards, same or worse results.

Execution is hard. I've always liked the phrase *practice makes permanent*. It's from my basketball and field hockey days. It's not glamorous; it is methodical, uncomfortable, and forces organizations to tackle decisions they have avoided for years. And the multiple steering groups that get set up for these practices? You need to find the ones who are ready to roll up their sleeves and get truly comfortable being uncomfortable. That is the only way to redesign workflows, decide which policies die and which live, agree why and how tech needs to change, and what needs to be measured and why it's important. This needs to happen in the daily operational systems of your business.

On the other hand, why it is easy to recognize failure is because it is usually in one of four places: people, process, policy, and technology or systems. So, I challenge you to ask, was the workflow designed to make it more convenient and easier internally or for customer ease? Were the rules designed to protect the business rather than enhance customer relationships? Were the changes made to help employees and teams deliver them or keep these folks constrained by the existing practices? And lastly, are we layering new tools on top of broken foundations and messy and dirty data? (What is that phrase? Garbage in, garbage out?)

To move from theory to practice, think about these observations. Clearly define who owns this transformation. Look at simplification across the organization and for your customers as the business rule. Empower your people to make good decisions and allow them to make the right one. And use technology that supports humans rather than trying to replace them. That is a great place to start.

You now believe in the Platinum Rule. You understand that culture is the operating system of your business. You recognize that leadership controls the temperature in every room. Now the real work begins to design a machine so powerful, so intentional, so human that caring for your customers and your employees is no longer optional, but unavoidable and inevitable.

CHAPTER 10:

POLICIES THAT DON'T SUCK

Most companies lose customers when they hide behind bad rules. Did you know that rules that are not customer-focused are *quietly destroying your customer loyalty*? Let's talk about what we need to do to fix that. If your customer service team's most utilized sentence is "That's our policy," and it destroys the customer experience, regularly ends conversations, frustrates customers, and shuts down solutions, you don't have a service strategy—you have a trust problem.

Let's be clear: Good policies are essential because they can help to create fairness, consistency, and clarity. And when done right, they can actually protect customers, support employees, reduce confusion, improve speed, and of course, build trust. Strong policies can make a business scalable by preventing chaos, ensuring equity, and explaining where the boundaries lie.

So, this chapter is not anti-policy; it's an argument against poorly designed, ineffective policies that fail both people and purpose. There is a difference. Somewhere along the way, many companies shifted policies into shields against customers, silencers on employees, and comfort blankets for leaders, fearful of risk and afraid of doing what is right.

Unfortunately, especially since Covid, policies started to eradicate judgment rather than strengthen them. They started safeguarding spreadsheets rather than relationships. They rewarded rigidity rather than accountability. And service ends when that happens.

I learned this the hard way as a loyal customer with a major U.S. airline.

I recently traveled from Fort Lauderdale to Philadelphia. Upon departure, I walked to the baggage claim, and there it was: My brand-new, $500 suitcase coming down the belt like it had been through a demolition derby. The zipper had exploded. Clothes were literally hanging out. The bag was crushed, bent, and scarred. There was no debate about what happened. Their equipment did this.

It was so badly damaged, I couldn't move it. In fact, there was no way to properly lift it on my own without leaving a nice trail of very pretty clothes and undergarments. I had a car service waiting for me to take me to my next destination. The airport was packed because it was just before Christmas. While I went to look for a representative at the airline's counter, the car service driver I had just met stayed with my ruined luggage. But there was no one there. My driver, Jack, told me it was his third airport run of the day—and his third time seeing a damaged suitcase, though none like mine. We both tried to come up with a solution while his idling car tried not to be towed. I'm going to lose my nice clothes, I thought. What else may have fallen out? Who is going to reimburse me for this damage? Should I call the airline? I felt humiliated, frustrated, and very tired.

After Jack dropped me off at my destination and I got my clothes inside, I decided to fill out a form online. What I faced was an illogical bureaucratic loop. It kept trying to make me take my bag back to the desk at the airport.

Obviously, I was now miles away with a suitcase that needed to be thrown away. When I called the airline, they asked me to fill out the form at the desk. I felt like singing, "Here we go 'round the mulberry bush..."

I spent another thirty minutes writing an escalation, received an email reply, and then went back and forth with the airline on that medium. Their final response was, "Unfortunately, our policy doesn't cover zipper damage." I was shocked. My zipper didn't just break. The bag was in beautiful condition before I left my prior destination—it had never been used! The airline destroyed it, let my belongings fall out and be exposed, and their response is a sentence from a rulebook?

My $500 bag turned into their $0 responsibility.

I pushed back. Politely at first. Then firmly.

Another sentence arrived. "Our policy requires the claim to be filed within X hours and must meet Y criteria." Translation: "We're sorry this happened, but we're more committed to the rule than to you."

I escalated. Supervisor. Manager. Same script. Same word. Same dead-end.

At no point did anyone ask, "What would be the right thing to do here?"

They weren't empowered to. They were trained to defend the policy, not the relationship.

And that's how companies lose loyal customers—over something that should have been an easy fix. This is what bad policy looks like in the real world. It turns obvious mistakes into emotional betrayals. It turns employees into human shields. It turns fixable problems into permanent damage.

Here's the truth leaders need to hear. Policies should exist to protect the customer experience, not destroy it. Great companies write policies that assume good intent, not abuse. They give employees discretion. They reward judgment. They understand that one loyal customer is worth far more than the cost of replacing a suitcase, refunding a fee, or bending a rule.

The moment your people are afraid to do the right thing because a policy might get them in trouble is the moment your system breaks. Policies don't scale trust. People do.

So, let's make this official today. Use policy for good. Replace it with: "Let me see what I can do." "I understand why you're upset." "We're going to fix this." "Here is what we can offer you."

Because customers don't stay loyal to rules. They stay loyal to how you make them feel when things go wrong. And nothing destroys that faster than hiding behind a sentence.

Why Policies Exist—and How We Broke Them

I know what broken policy feels like from the inside too.

Part of my role as an operational leader at Telefonica O2 UK was leading the escalations team. We handled both regular and executive complaints, the politically charged situations, the cases with potential to reach regulators. My team and I were exceptional at our jobs. We were named Escalations Team of the Year every year I was there. Internally, we were known as the people who could fix anything.

We were proud of that. And we should have been. But I came to understand something that took longer to admit than it should have: The existence of an award-winning escalations team is not a sign of organizational strength. It is a confession that the system above it is broken.

We were fighting every single day against policies that truly sucked.

This was the year other wireless carriers received the iPhone. In the UK, smartphones were no longer luxury items. They had become an extension of identity. Customers were no longer purchasing minutes and texts. They were purchasing access to their world and to the people and things that mattered most to them. And yet we continued to operate under policies designed for an era entirely foreign to the customers standing in front of us.

The upgrade policy was one of our worst battles. Loyal customers—some who had been with us for three, four, five years—would contact us expecting some recognition for their commitment. What they typically received instead were explanations of why they were not yet "qualified" to upgrade. Or why the phone they wanted would require a higher tariff. New customers off the street regularly received better deals than the people who had stayed with us for years. That always felt completely wrong. Because it was.

Then there were the early termination fees. Customers who needed to exit their contracts before the end date faced fees that could run into the hundreds or thousands of dollars. The formal justification was recovering the cost of the subsidized handset. What it actually did was make customers feel trapped.

I remember a father who had recently lost his job during the recession. He wasn't trying to avoid paying. He simply could not afford to. The policy contained no mechanism to assess whether a customer was in genuine financial difficulty. The fee was the fee. The representative could not waive it. So, it landed on us—the escalation team—to try to find a way through. Sometimes we succeeded. Most of the time we did not.

And then there were the return windows. A customer who discovered their coverage was inadequate, or that the device wasn't right for them, had a narrow window to act. Beyond the cooling off period, the answer was no. Not because anyone lacked concern. Because the policy said no.

By the time customers reached our team, they had already spoken to two or three other representatives. They were exhausted. They weren't calling about a phone anymore. They were calling about respect.

Here is the truth I carry from that experience: We were not the problem. We were leaders operating within the constraints of what were considered industry standards at the time. All carriers used similar policies. They were defensible on paper. Competitive by the metrics that mattered to finance. And they were quietly, systematically destroying trust.

When a loyal customer feels penalized for being loyal—when a parent in financial difficulty feels trapped rather than heard—the damage is not in the interaction. The damage is in the permanent message it sends: You stayed with us for five years and it means nothing. The customers stop advocating. They may not leave immediately. But they start waiting for the moment they can.

No escalations team, no matter how talented, can repair systemic indifference at scale. We could fix individual cases. We could not fix the policy that kept creating them.

What I learned is this: Policies that protect revenue without accounting for human dignity are not assets. They are liabilities that don't show up on any balance sheet until it is far too late. Loyalty is not compliance.

A customer who stays because leaving is too expensive is not a loyal customer—they are a hostage waiting for the exit.

You do not have to write weak policies to write fair ones. You do not have to sacrifice revenue to maintain trust. But you do have to be willing to ask, for every rule you write, whether the human being on the other end of it will feel seen—or whether they will simply feel processed.

We eventually did push back. The industry did change. But the learning cost us far more than it needed to.

THE TEMPTATION TO TURN POLICIES INTO SHIELDS

Policies were originally created for good reasons: to provide consistency, fairness, and protection. Somewhere along the way, companies twisted them into something else: a shield against customers and a muzzle on employees.

Instead of empowering people to do the right thing, policies now often remove judgment, they punish initiative, they reward rigidity, and protect the company at the expense of trust. And when that happens, service dies.

In actuality, your policies don't matter to your customers. They are interested in your decisions. Whether you resolve the issue is important to them. Whether you listen is important to them. When anything goes wrong, people are concerned about whether they feel valued. No client has ever remained faithful because of a policy. They stick around because of how you made them feel during a significant time.

In Chapter 9, we talked about execution, the mechanics of service, systems, workflows, decisions, and behaviors. Now we know that policies are a core part of your operating system. They silently tell your employees what matters, what is safe, and what will get them in trouble. When your policy says, "don't make exceptions," your people stop caring. When your policy says, "protect the company first," your customers feel it. When your policy punishes initiative, your culture decays. You cannot claim to care about customers while writing rules that actively prevent your people from helping them.

Great policies protect the experience not just for the company. They trust employees to use judgment, and they assume good intent from customers, not abuse.

The best organizations understand that flexibility is not weakness, it's intelligence. They build guiderails, not jails. They train people on principles, not just procedures. They recognize that loyalty is earned in the exceptions, not the rules.

The Real Cost of Bad Policy

That airline may have saved the cost of replacing my suitcase. But they lost far more in trust, advocacy, and future revenue. Bad policies don't show up on balance sheets. They show up in churn, negative reviews, and the slow erosion of your brand. And once trust is gone, no amount of marketing can buy it back.

The Other Side of the Coin: Policy Done Right

To finish the airline story... I had to get a new suitcase. As you already know, I'm a world-class shopper—if shopping were an Olympic sport, I'd have a medal. While heading to my next destination, I realized I needed the bag I trust most. I usually stick to their carry-ons because they're sturdy but heavy, and their larger suitcases can add too much weight. Still, this time I needed something durable—something that wouldn't break. So I did what any good shopper would do. I started calling stores along the way to see who had it in stock. Amazon had it, but delivery wasn't guaranteed before my departure, especially with a snowstorm coming. The only reliable option was to pick it up myself. Then I saw the brand was having a sale. I called the one store I was sure would have it—and they did. Thank you, Nordstrom. I called the luggage department—not always an easy task, since those calls usually go unanswered—but this time, someone picked up right away. She was pleasant and helpful. I explained that the website showed one suitcase in stock and asked if she could confirm before I made the trip. She put me on hold, checked, and came back to say she had found a brand-new one in the back and would set it aside with my name on it. I was honestly shocked. It was so easy, so smooth—such a relief.

The next day, I headed to Nordstrom just before a major snowstorm was set to hit. The store was packed, and the luggage section—smartly located within handbags—was full of holiday shoppers. When I arrived, the same woman greeted me, said she'd grab my bag, and returned shortly after. Again, I felt relieved knowing I'd found a durable replacement to match the many pieces I already had waiting for me back home, nearly 3,000 miles away.

At checkout, I mentioned, "I'm not sure if you price match, but the brand's website has this on sale for a lower price." She immediately said she could match it. Then, without hesitation, she added, "Don't worry about it." She didn't ask for proof, a screenshot, or anything.

Just like that—relief and surprise all over again.

She barely glanced at my phone. She didn't ask for proof, didn't mention needing approval from a manager, and didn't even explain the store's price-matching policy—she just did the right thing. She trusted me. It was incredible.

I thanked her profusely and wheeled my luggage out of the store, genuinely impressed. She took initiative, acted with good judgment, and clearly felt empowered to make the right call. In that single interaction, she created a lasting sense of loyalty for me toward Nordstrom.

Nordstrom consistently seems to get customer care right. Its return policy, often described as legendary, has long been associated with flexibility and generosity—famously illustrated by stories like the decades-old tire return. While the policy has evolved over time and is now handled more on a case-by-case basis, the core philosophy remains the same: Prioritize customer satisfaction and experience above all else.

You see, it wasn't so much the policies that built loyalty, but the people. If your policies make it harder for employees to help customers, those policies don't just suck, they are rapidly destroying your business and your customers' loyalty. And the fix doesn't start with rewriting the rulebook. It starts with leadership deciding. And the belief from customers is that Nordstrom is a company that makes things right. They seem to believe their customers, trust their people, and care more about the relationship than the rule.

That's what customers remember. That's what brands are built on.

Amazon does exactly the same. If something is wrong, they fix it, fast. Amazon also uses policy proactively for the customer experience. They built their brand on this as well. If something is broken, missing, and not as expected, they refund you instantly without interrogation, without paperwork theatrics. They demonstrate trust, empowering their associates to do the right thing, and prioritize speed, trust, and scale over continuous layers of approvals.

So, if your policies are designed for the convenience of your legal team, your finance department, or your systems administrator, and not for your customers, you will continue to have a loyalty problem. If your policies help empower your employees and make it easier for the customer when something goes wrong, you will win.

Too many companies remain paralyzed by policy. What begins as a reasonable effort to create consistency slowly mutates into a bureaucratic nightmare. A refund that should take two minutes takes two weeks. A simple password reset requires three approvals. A loyal customer with a spotless history feels treated like a potential criminal.

For one, I stopped arguing—I leave. Most customers will do the same.

The Bottom Line

If you want customer loyalty and a better customer experience, if you want to demonstrate to your customers that you really care, make it easy for them to transact and stay. Make it easy for them to be heard. Make it easy to do business with you, and make it easy to fix mistakes. Stop hiding behind rules that drive customers away.

Your policies shouldn't suck. They should demonstrate they do the right thing and that they are fair and can protect the business in the right way. Policies should be human, fair, flexible, protective, and empowering.

Policies are not bad. They must never come at the cost of employees who lack the permission, the power, and the courage to fix them, and customers who need to know you will do the right thing. Policies are supposed to protect your business, not destroy your customer relationships. Policies must come with humanity.

Stop shrinking behind policies that try to protect your business but instead punish your customers and drive them to your nearest competitor.

Leadership must decide that relationships are more important than regulations. Customers remember that. This is how brands are built and what keeps growth going. Change your operating system today.

CHAPTER 11:

OWN YOUR MISTAKES AND FIX THEM

Let's face the facts. Your customers don't expect you to be perfect. They expect you to be accountable. Mistakes happen all the time, and things do go wrong. Systems fail. Packages get lost. Flights get delayed, websites crash, online orders go sideways, and sometimes emails don't actually go through. And the sad reality is that we all know we'll never live in a perfectly constructed world. Mistakes will always happen, even with technology... especially with technology...

The fact that mistakes and system failures happen does not surprise customers. It is not the mistake itself that angers customers, but what follows after. It is the silence. The blame game. The deflection. The lack of resolution. And we all know how policy quotes make customers leave.

For the last several decades, I've told my teams that it's not the mistake that creates anger in customers but how we correct those mistakes. In the customer experience field, how you react when things go wrong says volumes about how much you care about your business and what your brand represents. And boy do I have a story for you. But first, isn't it surprising how many companies refuse to apologize?

The Power of an Honest Apology

An honest apology is one of the least utilized tools available to your organization in the area of customer experience. I'm not referring to the generic, canned, lawyer-approved response. I recall working for a company some years ago, when our chief legal officer instructed my team "never" to admit a mistake. I mean, we were selling mobile phones, not vaccines. She told us that admitting a mistake admitted liability. Absolutely not.

Saying "We're sorry for any inconvenience this may have caused," is completely different from saying, "We messed up. Here's what happened. Here's what we are doing to correct it." These sentences, delivered sincerely and promptly by humans, save more customer relationships than any loyalty program ever will. I guarantee it.

One story I have told for years illustrates why I have had eight Lexus RX vehicles over the last twenty-four years.

The Car That Smelled Like Death and the Brand That Earned My Loyalty Forever

I had such intense brand loyalty to Lexus and to *my* Lexus dealership that when I moved from the north of the UK to the south for a new role, I actually called my dealership to say goodbye.

I know this is not typical behavior. However, to me, Lexus was more than just a car brand; it was a relationship. I purchased my first Lexus RX 300 from this dealership, and I brought my car there for service. I would often take over one of their offices and quietly work there while waiting for my service appointment to end. I truly felt like I was part of their team. So of course I called them to say I was leaving the area.

"Funny you should call," the manager said. "We have a new car you might like." My lease wasn't up for another year, so I laughed and said, "Not unless it's burgundy with beige leather." Without missing a beat, he replied, "That's exactly what it is."

I paused for a moment, and then I said, "I'm moving *tomorrow*."

"No problem," he told me. "We'll drive it out to you and collect your current car. We'll waive the down payment and lower your monthly payment."

At this point you're probably thinking: *This woman is a marketer's dream.*

He told me to take my time, and I told him I'd consider it. He assured me they wouldn't sell the vehicle to someone else but would reserve it for me. I called him the following week: "Let's do it." I didn't buy the new car because I needed a new vehicle—I wanted the new car, it was a better deal, and it was literally the exact color combination of my dreams. For the first four months, I loved it. I even purchased a purse and a beautiful pair of leather pumps to match, and also a few new business suits. I mean, of course I did.

Then, about six months later, the smell started.

Each time I started the vehicle, a wave of something so disgusting, so awful, so obviously terrible came rushing through the vents. I was convinced an animal had died in my engine or someone was intentionally poisoning me.

I took this beauty to my local Lexus dealership. They cleaned the air filters, changed the settings, ran ten different diagnostic tests, and declared the vehicle fixed. Two weeks later, the odor returned.

Back I went. More tests. More repairs. Losing hope and confidence slowly. And then the odor returned... again.

Each time I turned on the air conditioning, I felt like I was walking into a chemical warfare facility. By that point, I decided to escalate the issue further. Not to the CEO, nor to the COO, nor to the head of customer care, but to the Lexus UK escalations department.

They sent me an email: "We are sorry. Someone will contact you soon."

The following day, they called me: "If we cannot repair your vehicle, we will replace it." I experienced the entire gamut of emotions: I was in disbelief about what an escalations agent said. That type of confidence and direct ownership of the problem floored me. Then my thoughts started racing. Why hadn't they fixed the vehicle earlier? Had my trusted dealership sold me a defective car? I felt overwhelmed—how could this be happening? Had anyone else noticed the odor? What if they couldn't fix it? Was I going to lose my car?

During the conversation, I was silent. I didn't utter a word. I simply stared at my phone like it had spoken a language I was not used to hearing or understanding.

Eventually, they picked up my vehicle, provided me with a brand-new Lexus loaner, and told me they would investigate.

Two days later, the escalation agent assigned to my case called. She said, "I'm sorry to inform you that we're unable to repair your vehicle. Please let us know which model you would like as a replacement."

It takes a lot to leave me speechless—but in that moment, I was. I was still in shock. I instantly cycled through every stage of grief: denial, anger, bargaining, depression, and finally, acceptance. For a moment, I even considered switching brands. I asked if I could see Sheila—the name I had given my car—one last time, but they told me they would be holding the vehicle until my replacement arrived.

I started thinking about the insurance I had paid and the tax costs I had covered over the past year for a vehicle that ultimately failed. "We'll reimburse you for that," she said.

That was the moment I had to sit down. I had gone into this expecting the worst. I was prepared for frustration, delays, and a long, difficult process—something I would simply have to endure. So, when Lexus exceeded every expectation I had, I was almost unprepared for how seamless and supportive the experience felt.

Now eleven years later, I'm driving my eighth Lexus RX. Not because the vehicle was perfect—but because when it wasn't, the company was. They didn't deflect, make excuses, or hide behind policy. Instead, they turned a failure into a lifetime of loyalty.

To this day, I've remained a committed and enthusiastic Lexus customer—for over twenty-five years. Southwest Airlines learned the same lesson in December 2022. During what should have been the busiest travel week of the year, a winter storm collided with an aging scheduling system that simply could not cope. More than 16,000 flights were canceled over ten days. Two million passengers were stranded. It became the worst operational meltdown in commercial aviation history.

Here is what Southwest did next: They owned it. CEO Bob Jordan issued a direct video apology. No corporate hedging. No deflecting to the weather. They committed to refunding every affected ticket and reimbursing out-of-pocket expenses. They publicly committed to overhauling the scheduling systems that had failed. The Department of Transportation investigated and levied the largest consumer protection penalty in aviation history. Southwest paid it.

Was it perfect? No. Was it enough for every stranded passenger? Probably not. But the transparency mattered. When a company stands in the fire rather than hiding behind fine print and force majeure clauses, customers notice. Trust, once cracked, can be rebuilt. But only if you show them you are willing to pay the cost of the mistake you made.

Chipotle displayed that same bravery in 2015 and 2016, but with far higher stakes. A series of E. coli and norovirus outbreaks struck their restaurants across the country. Customers got sick. The media descended. Same-store sales dropped by nearly 30 percent. It was an existential moment for the brand. What made Chipotle's response remarkable was not just speed—it was depth. They did not quietly retrain a few managers and hope the story would fade. They publicly acknowledged what had failed, voluntarily closed locations for safety audits, redesigned food handling protocols from the ground up, and brought in dedicated food safety leadership rather than folding the responsibility into an existing role. Radical transparency, executed at real cost, became the foundation of a second chapter. Within a few years, the stock had recovered. By the early 2020s, Chipotle had become one of the most valuable restaurant brands in the world.

Do you remember the airline that experienced a spectacular failure? In 2017, it forcibly removed a paying passenger and physically dragged him from the plane while other passengers filmed in disbelief. Instead of immediately accepting responsibility, the airline blamed the passenger and cited policy. The delayed and insincere responses to the incident sparked a PR disaster, generated massive backlash, and severely damaged the brand's trust with its customers. And they demonstrated that the longer you delay accepting a mistake, the faster the story will spiral beyond your ability to manage it.

A failure when properly handled creates opportunities for enhanced loyalty. A customer who witnesses you acting quickly, empathetically, transparently, and responsibly in the face of adversity will be far more trusting of you than one who has never faced an adverse situation. That is not a theoretical concept. It is the essence of humanity.

Memorable experiences for customers are not costly. They are human. A hotel provides a complimentary room to a guest after it overbooks the rooms. A subscription service credits a subscriber's account for the month after a billing error occurs. An agent writes a handwritten letter to a customer after resolving a significant complaint. A manager personally calls a customer after a delivery fails. The stories are shared. They are the basis upon which brands are founded. That is the point at which intention transforms into execution.

For your company to handle errors in a productive manner, it is necessary to create a strategic plan. Your employees need to understand the protocols that guide how to respond to failures: acknowledge them, accept responsibility for them, resolve them, and follow up. No sixty-six approvals required. No policy paralysis allowed. No evading through process.

Educate your employees on the characteristics of a true apology. Practice the act. Reward the behavior. Identify your primary complaints and resolve their underlying causes, rather than just treating the symptoms. Respond to an issue both in public and privately. Silence breeds anger. Transparency generates loyalty. When an employee diverges from the script to correct a situation, recognize and encourage the behavior until it is an integral aspect of your corporate culture.

You will make mistakes. Many of them. Every company does. The only question is whether you will respond as a company that shows that you care and are concerned or as a company that conceals those concerns.

Error is inevitable. Trust is discretionary. The difference between excellent companies and those that are forgotten lie not in the number of failures they experience, but in how seriously and how committed they are to recovering from those failures. The speed with which they accept responsibility for a failure, how clearly they accept that responsibility, and how personally involved you are in correcting the failure determine how well they recover. Your reputation depends on you handling this correctly. Technology cannot do everything for you. People will. In our boardrooms and budget meetings, we continue to pretend otherwise.

CHAPTER 12:

TECHNOLOGY IS NOT A REPLACEMENT FOR CARING

We live in an age of rapid technological advancements that have transformed the way businesses create value by replacing people with machines, transferring responsibility for accountability from humans to machines, and creating experiences that are convenient but lack a human touch—and then calling it innovation. We celebrate the "speed" of innovation over its sincerity; the "efficiency" of technology over the experience of being human; and the ability to scale, i.e., reach as many customers as possible, above all else. Yet regardless of how sophisticated or advanced our tools become, no chatbot will ever be able to understand the pain of disappointment, no dashboard will ever be able to identify and respect dignity, and no AI model will ever be able to restore a damaged relationship.

Technology, however, can support service, strengthen operations, and simplify lives, but it can never replace caring. When organizations forget that, they don't just lose customers. They lose trust, relevance, and the very humanity that once made them worth choosing. You will hear more about this in some of the later chapters and its impact on brands.

Automation and AI should amplify our humanity, not replace us. AI is on fire. Automation has gone wild. Self-service is brilliant. But, you need to understand—technology does NOT fix a bad customer experience; in fact, when it is not done correctly, technology can take a poor customer experience, create it at scale, and make it more annoying for your customers as they are served by more technology.

I have sat on both ends of the table as technology has changed how we provide customer experiences. Systems I helped design, I watched fail from their own ambition. My first lesson on using automated caller recognition and not first purging the data years ago was one of those lessons. On paper, the call flow and the technical architect diagram were beautiful.

The system could recognize the customer's number and pull up their information and route them to the correct team, while calling them by name. But I didn't realize until after that there was a lot of disorganization with the customer information, including a large amount of missing data in customer profiles. Many of the customer numbers were also incorrect, and many customers had multiple phone numbers they wanted to use. The main issue was no one provided the system with clear direction on how to act in each of the possible scenarios that may occur. Therefore, when the actual situation deviated from the original design, the system continued to make errors over and over again.

As you can see from my experiences, technology is good. Technology does not judge, it does not have feelings, nor will it ask you a question. Technology will operate exactly as you programmed it to. The only thing that determines how well a system works is how well it has been designed. When someone says, "the system is operating as it was designed," you know that either the system has done an excellent job or a poor job of doing whatever it was intended to do, depending upon the intentions of those who designed it. Therefore, it is crucially important to ensure that your technology is enhancing the overall customer and employee experience and not diminishing it.

I learned this at a much more painful scale when I became one of the first leaders to bring real-time voice translation into a full production environment. The goal was genuinely exciting: remove language barriers, open new markets regardless of geography, and build an operation that was more agile and less constrained by international borders. I believed in the vision completely. In retrospect, I may have believed in it faster than the ecosystem around me was prepared to absorb.

I did not undervalue the technology. I underestimated the human side of the system in which that technology had to live. My organization was not prepared. The product team had concerns about the maturity of the feature. The technology team raised flags about reliability and reputational risk. Operations worried about how customers would react to speaking with an agent whose voice was being translated in real time. There was limited to no buy-in—not because the idea was wrong, but because I had not built the internal coalition necessary to carry it. The partner providing the technology was also learning in real time. They told us no integration would be required. That was not completely true. Their platform was not yet ready for the customer privacy and data requirements our environment demanded. And while the technology translated words accurately, it could not always reproduce the pace, tone, or cultural nuance of a native-speaking representative. Those edge cases exposed gaps. Confidence inside the organization eroded quickly.

There were people celebrating what we were attempting. There were also people convinced I was simply trying to eliminate agents and reduce headcount. I was not—but I had not done enough to make that clear, and once an organization loses faith in a pilot, enthusiasm for expanding it is nearly impossible to rebuild. We stabilized the experience. We improved the technology. We demonstrated that the concept worked. But by the time we had, the internal trust capital was spent. The ambition to expand into additional languages and create a genuine competitive advantage was halted—not because the technology failed, but because the humans around it had stopped believing.

The lesson I carried forward had nothing to do with voice translation. It had to do with change management. Organizations do not want a 90 percent solution. They want certainty, and when they don't get it, even a genuinely good idea becomes the thing they tried once and decided not to try again. Human trust builds over time. Technology develops very rapidly. When those two speeds are not managed deliberately, even the most innovative ideas will struggle to survive long enough to prove themselves.

If I were to do it again, I would allocate as many resources to building the internal coalition as I allocated to evaluating the platform. I would over-invest in simulation, in running scenarios with frontline employees, in setting expectations clearly before the first day of the pilot. I would make sure leadership was genuinely behind it—not neutral, not watching to see what happened, but invested. Because I learned this the hard way: If your leader is not supporting the effort, you have no chance of changing hearts and minds. My leader was on the fence. No passion for trying it, and no passion for it succeeding. That fence is where innovation quietly dies.

The voice translation experience taught me about moving faster than the humans around me were ready to absorb. A later experience as chief operating officer of a large home services company taught me something different—and in some ways more unsettling. What happens when you do everything right, and the system still surprises you?

We were not simply following a trend. We wanted to demonstrate our commitment to lead responsibly in a rapidly evolving world where automation was becoming increasingly prevalent. Customers were demanding quicker solutions. Labor costs continued to rise. The mandate was clear: Design an operating model using automation to increase productivity, reduce labor costs, and resolve customer issues more quickly—and do so in a way that elevated the human experience.

We were deliberate. Thoughtful. Systematic. Working groups were formed. Tiger teams. Focus groups with frontline agents. Senior executive steering committees. Journey maps were developed. Potential scenarios were white-boarded. Edge cases were identified. Escalation triggers were debated. Ownership rules were established. Routing logic was developed.

The model we built was elegant. Automation would handle routine work and serve as the first point of triage. Customers would be correctly routed to the appropriate team the first time. If they required more complex assistance, they would be transferred to one designated case manager who would own the case from beginning to end. No more bouncing between departments. No more repeating the same story to three different people. Complex customers would receive concierge service. One human. One relationship. One owner. On paper, it was beautiful.

We hired for it. New roles were created consistent with the model. Skill gaps were addressed where technology alone could not fill them. Staffing levels were set to projected volumes. We genuinely believed we had mapped the necessary use cases and edge cases.

And for a period, it worked. Customer satisfaction scores improved. Resolution quality went up. Escalation rates came down.

Then we discovered what we had not anticipated.

When automation removed the simple work, what remained was not proportionally distributed. We had assumed something like 70 percent simple cases and 30 percent complex. What we got was a dramatic skew toward the extreme end of complexity. We had unintentionally created a concentration of cognitive load. Case managers were no longer balancing simple and complex interactions throughout their day. They were handling only the hard ones. Emotionally charged cases. Cases involving sensitive financial information. Cases spanning multiple touch points, multiple parties, regulatory complexities, exceptions, gray areas.

Only so many of those can a person handle at a given time. And when a case manager was unavailable—vacation, illness, training—the volume did not disappear. It was absorbed by another case manager who was already carrying their own full load of complexity. What had been designed as a strength—dedicated ownership—had quietly become a fragility.

Automation does not eliminate work. It changes it. And sometimes it magnifies it.

We had not miscalculated the amount of work. We had miscalculated how the work would redistribute itself. To maintain the same quality of customer experience, we needed significantly more highly skilled people than our model had indicated. The cost curve shifted upward. The economics were not what we had planned for.

Our operational model was functioning exactly as designed. The design just hadn't predicted what complexity looks like when you remove everything that isn't complex.

What I learned from this is that automation is powerful, necessary, and capable of dramatically elevating the customer experience. But it changes human work in nonlinear ways. When you automate the simple, what you leave behind for your people is harder, not easier. If you do not account for that shift in your workforce planning, your capacity models, and your investment in employee support, you will either burn out your team or fall short of the commitments you made to your customers.

The trade-offs we faced were not failures. They were leadership decisions. Cost efficiency versus continuity of ownership. Scalability versus personalization. Automation benefits versus the cognitive load carried by the humans left in the system. Short-term margin versus long-term trust equity. Being ahead of the curve does not exempt an organization from unanticipated consequences. It just means you encounter them first.

Plan for the intensity of the work that remains, not just the volume of it. That is the lesson I would not trade.

Technology is meant to improve the lives of both employees and customers. Automation has the capability to assist in completing transactions quickly and efficiently, complete repetitive tasks, allow for faster resolution, minimize barriers/friction, and allow your employees to focus on the things that truly matter and/or the most difficult transaction/emotional issues that your employees are best suited to handle at least for now, such as thinking, caring, solving complex problems, and using their imagination/thinking creatively outside of the box.

When technology is developed or used improperly, it can become a significant hindrance to the overall experience. Examples include the chatbot that continues to loop and provide the same limited options regardless of how thoroughly you describe the issue/problem.

One recent example was when I called my mother's electric and gas provider to report she had no hot water. The company's automated system couldn't handle my requests or questions. After twenty-five minutes of looping through menus and getting nowhere, I finally opted to wait for a live agent—another fifty minutes later.

When I eventually reached someone, she was able to resolve everything quickly. Before ending the call, I asked her to please pass along feedback to management—that the automation system was failing customers and driving unnecessary calls to agents. Without hesitation, she laughed and said, "We tell them that all the time—they just don't care."

That response said everything.

Another example is the "Do Not Reply" email from customer service, which sends the exact opposite message of what customer care should represent. And then there are IVR systems designed to reduce handle time, yet they often increase the total time customers spend on their phone by fifteen to thirty minutes—forcing them to navigate frustrating, inefficient processes just to get help.

As mentioned previously, improper development and use of technology can result in a completely opposite outcome from what was envisioned. While you may be saving money and controlling costs, you are creating a blunt instrument to cut costs at the expense of loyalty. Customers can instantly determine if your technology was designed to empower them or deflect them. Once customers determine the latter, trust is lost.

Unfortunately, far too many organizations view technology as a shield, rather than a bridge. Instead of increasing the level of service to customers, the technology is designed to create obstacles that frustrate and prevent resolution.

I once worked for an organization that launched a new AI chatbot designed to "handle tier-one support." On launch day, customers realized that typing "agent," "representative," etc., simply returned them to the same loop of pre-defined responses. There was no exit, no escalation, no human interaction. Social media erupted. Customers stopped using the chat feature. Agents received a deluge of angry calls. The organization invested hundreds of thousands of dollars developing a system that made service harder to access. You've likely encountered these common culprits.

Each of the examples listed above represents a technological approach that is primarily focused on improving the operational efficiencies of the organization versus the customer experience. Reducing the time spent by agents on handling customer calls means little, if not nothing, if it results in increasing the days (and potentially weeks) of frustration that customers experience in resolving their issue.

The solution is not to eliminate automation. It is to develop automation that serves people, not avoids them. I'm saying this repeatedly because to me it appears to be a simple concept, yet unfortunately, numerous organizations continue to design their technology to achieve the opposite of this objective, as late as 2026.

It's funny to me that many organizations view eliminating an opportunity to interact with their customers over the phone or via voice call as a significant advantage to themselves as an organization. Definitely not an advantage to their customers. I love to break things down into two categories: "wanted" calls and "unwanted" calls. As long as technology can provide a solution to all those types of "wanted" calls, such as "pay my bill," "send me a copy of my bill," "where is my order," etc., I would be fine with completely getting rid of them.

However, when I want to ask a question that requires some complexity, is based upon how I feel, or is based upon a detail that I am trying to get out of someone, please make it a "wanted" call and service the call accordingly. Please make it easy for me to speak with you for whatever reason I called. If I can easily find my own answer using self-service options, then please allow me to do so. If you truly care about the journey or the experience, you can design for it, test it, and execute it.

American Express has done a tremendous job of identifying intent, recognizing repeat customers, and routing appropriately. They have also been successful in providing timely resolutions to my questions and issues.

You hear the term "personalization" being used by almost every executive team. You see, personalization does not mean simply putting a person's name in the subject of the email or using the person's name during a call. Personalization means that the organization has remembered YOU, understands YOU, and is able to modify the experience to meet YOUR needs in a manner that supports YOU.

Spotify does not simply tell you that we value your data. Spotify uses it to surprise and delight. Every year Spotify produces an amazing campaign called "Spotify Wrapped," which converts each user's music listening patterns into a story that the user shares with others. That type of personalization creates an emotional bond with the user.

I began my career in learning and development. I feel fortunate to have begun there because learning and development will always be the foundation upon which I think and the way in which I communicate. Mastering the ability to teach and make it easily understood and absorbed is significantly more difficult than the actual learning process. I bring this up because we were always taught that true success lies within connecting with the customer on many different levels. Therefore, if you can connect your customers to the technology, automation, and AI through a variety of channels, then that is truly a win.

As I explained above, when something goes wrong, as was the case with both my Lexus and airline baggage examples, customers want to deal with a human who can listen, empathize, accept responsibility and take ownership, and provide an appropriate course of action or solution. The airline did everything they could to provide me with their solution; however, they continued to send me multiple emails from multiple representatives asking for the same information and failed to read what I had previously provided to them. This is a huge difference between the experience of feeling seen and heard versus feeling extremely frustrated and dissatisfied.

Your technology implementation should make it easier for humans to appear sooner rather than later, be better informed, and be even more empowered to solve the customer's issues instead of making it harder for humans to be reached or to never be reachable at all.

While using technology to quickly solve a customer's problem is a positive, technology that enables you to quickly and efficiently resolve the customer's problem is equally impressive. In addition to being an expert shopper, I have also mastered the DoorDash method of ordering food. Although things can go wrong, DoorDash has successfully made it easy for me to obtain resolution in their application as opposed to having to speak to someone. Their application makes it simple, quick, and usually resolves the issue at hand.

In my opinion, success should be evaluated differently than it is today based on whether the technology implemented or the changes implemented reduce customer complaints and/or escalations as opposed to reducing telephone calls.

Additionally, simply monitoring "containment" rates in self-service is not sufficient. For example, if 50 percent of customers attempt to resolve their issues on their own, but 50 percent of those customers ultimately call back to have the issue resolved, that is not efficiency—that is failure. Real success is defined by problems being resolved the first time with no subsequent callback(s) and/or second stage complaint(s).

Instead of providing a "cold transfer," design a "warm transfer" for escalations. When a customer reaches the limit of self-service, they should be transferred to a human without interruption. Include a copy of the transcript, their identity, and their intent so the agent can continue immediately where the bot left off and without restarting or repeating.

Those organizations that design this well inside the automation and AI they are developing are going to be the real winners. I witnessed this in a call regarding an online Walmart order. My returns process was exceptional, and the AI seamlessly transferred me to an agent when it could not offer me the resolution needed, and the agent did so in under ten seconds. It was one of the best examples I have ever seen of AI handling the simple and an agent handling the complex, with AI making the decision and without me repeating myself eighty times, yelling at the phone, or having to speak louder. Well done!

We collect so much data and insight regarding how customers utilize language, and we should be utilizing this insight to make it much simpler for businesses to achieve these types of transitions. For example, we already know that you are calling back a second, third, fourth time from the same number on the same issue.

When your service is so good that customers rarely have to ask for anything, you've done something special. Amazon has done an incredible job of being proactive with their technology (not reactive). Amazon lets their customers know if their shipments will be delayed or if there's an issue with receiving their product, so when utility companies let you know about planned service disruptions, or retailers send you back-in-stock alerts, you're doing the same thing: enhancing the customer experience by using technology proactively. When you tell your customers about an issue before they see it themselves, you build trust with your customers and eliminate anxiety. That's 100 percent true.

The companies that do this successfully are using technology to augment their humanity, not replace it. Automation should increase empathy, speed, and clarity, not diminish these attributes. Your customers won't abandon you because you're embracing technology. They'll abandon you because the technology you chose made them feel ignored and invaluable.

How many times have we seen in the news this past year that companies have used AI to cut costs, rather than saying that they have improved customer experiences or customer moments? How many times have we seen companies define their successes solely through deflection, reduction, and elimination, rather than resolution and solved problems?

One of the most critical decisions a company can make is to choose its appropriate technology stack. However, far too often, the selection of a technology stack is treated as a procurement decision rather than a strategic one. A chatbot here, a CRM upgrade there, a new ticketing system there. We often purchase solutions in a vacuum, buying a solution without first establishing the experience we want to deliver to our customers, and that is the biggest mistake companies make.

If the primary purpose of technology is to solve internal problems—rather than customer problems—the end results are usually disjointed systems, confused employees, excessive amounts of unnecessary data, and customers who find fault with each interaction. For both humans and machines to respond intelligently and consistently, a modern technology stack should serve as a single, cohesive nervous system that allows customer identity, history, preferences, and intent to flow seamlessly across multiple channels.

To evaluate a technology stack, companies must begin asking the right questions. Does this technology make the customers' work easier, or simply transfers the burden from us to them? Does this platform provide additional displays and regulatory requirements, or front-line teams with more context and control? Will this technology create additional silos, or allow seamless integration into existing systems? Does this technology only work when everything is working perfectly, or can it adapt to the realities of "real-world" scenarios? The best technology stacks are the ones that are the most flexible and can handle the complexities of humanity, unexpected events, and poor-quality data, not the ones with the most features.

Technology success happens when customers are able to self-serve for the basic and heavily transactional items (e.g., billing history, order edits, tracking, etc.). Utilizing AI to triage is spot-on and assists in directing customers to the next available resource, team, or human much faster. Providing proactive notification to customers generates vast amounts of loyalty and trust, while utilizing agent assist tools to enable front-line employees to answer customer inquiries more efficiently and effectively reduces the amount of time wasted by customers.

Poorly executed technology can erode trust with customers when they cannot easily contact a human, when bots refuse to refund a customer, or when a customer receives an apology from a machine (i.e., the system) instead of a human. Customers do not leave companies because of the technology they employed; they leave due to how the technology made them feel or did not feel and how much effort it took to transact.

Making customers repeat themselves over every channel is a massive failure. If you are not passing information along, you are not understanding how to utilize your technology tools to empower humanity. If you develop a bot that does not recognize when it can no longer assist, that is also a massive failure.

A technology stack should be evaluated based upon how regularly it creates confidence in your customer experience and how consistently it empowers your employees, not based on how impressive it may appear during a demo. Regardless of how sophisticated it claims to be, it is not a suitable stack if it does not positively impact both areas.

Technology is not the enemy. It is a tool. Like any tool, the intention and capability of the person operating the tool will ultimately determine the effectiveness of the tool.

Part IV–

THE BUSINESS CASE FOR CARING

This is the section where conviction becomes validation. Where values become valuation. Where faith becomes concrete financial reality.

CHAPTER 13:

THE ROI OF DECENCY AND DOING THE RIGHT THING

By now, I hope you've been convinced that caring is not some sentimental weakness. Culture is required for any company. People are not the "nice-to-have" element of an enterprise; people are the operating systems by which all successful and enduring enterprises operate. Faith is converted into evidence. Intent can now be quantified. Values appear on the balance sheet.

Caring can be measured, validated, and replicated. When used with intentionality and purpose, caring can be one of the most predictable and replicable means of expanding an organization today. Organizations that provide better treatment to their customers and employees will continue to generate higher revenue, better employee retention, stronger brand reputation, greater organizational resilience, and more sustainable performance. Loyalty builds upon itself. Risk decreases. Brands endure.

Compassion is an asset. It's a competitive advantage, a leadership skill. The business case for caring proves that doing the right thing is not only the correct moral decision—it is also a fiscally astute one. Profit and empathy are not mutually exclusive. They are complementary. They have a symbiotic relationship.

The world's most successful, prosperous, and profitable companies have discovered a simple yet powerful formula: Designing for people leads to natural performance. Customers spend more, stay longer, forgive more quickly, and speak out more. Employees are more committed, engaged, and productive. Turnover decreases. Costs stabilize. Momentum builds. Caring permeates throughout an entire system when it is integrated into that system.

For years, business executives have spoken about employee engagement, culture, and customer experience as though they were sensitive topics; important, yes, but secondary to "true" business concerns. Those days are behind us. Compassion is now measured, monitored, and benchmarked. It has a direct correlation to earnings, growth, and resilience. The data is clear: Organizations willing to invest in the way people feel and solving problems will consistently outperform those unwilling to. Compassion has become one of the most reliable financial engines an organization can develop.

This is where the importance of return on impact (ROI) is clear. While most organizations focus on ROI—"What did we spend, and what did we get back?"—very few organizations rigorously monitor return on impact, ROI, the emotional, relational, cultural, and reputational impact of the decision(s) made on employees and customers." How often do you ask yourself, "What is the return on how we treat people?" "What is the return on our culture?" "What is the return on trust?" "What is the return on dignity?" Do you evaluate those returns with the same level of detail as you evaluate margin and cost structure? If you do not evaluate the return on impact of your decisions, the return on investment of those decisions will eventually collapse.

One of the most obvious examples of return on impact is employee engagement. Gallup's independent research has shown that companies with high levels of employee engagement outperform their peers by as much as 147 percent in earnings per share. Regardless of economic cycles, companies that are engaged are able to recover more quickly, grow stronger, and adapt more rapidly than their competitors. This is not coincidence; this is cause and effect. Engaged employees think differently. They fix problems rather than simply pass them along. They protect customers rather than simply processing them. They innovate rather than simply react. They care because they feel cared for.

There is no number of perks, slogans, or quarterly town hall meetings that can generate this type of engagement. It is based solely on the consistency of leader behavior. It is based on being heard, trusted, and feeling connected to the company's mission and purpose.

When people feel that their work matters and their voices matter, performance improves. Productivity improves. Errors decrease. Turnover drops. Institutional knowledge increases. Service is more consistent. Engagement creates an operational advantage.

Customer experience follows the same model. Research from Forrester and other analysts show that exceptional customer experiences drive loyalty, repeat purchases, and advocacy, three of the most effective revenue multipliers in business. Customers who feel valued and supported purchase more, retain longer, complain less, and advocate more. This is not branding. This is economics. Experience is revenue delayed.

Engagement is a total driver of performance and not simply a "feel good" or "soft" element. Gallup's large-scale workplace studies show that teams in the top quartile of engagement perform better across all areas of profit, productivity, retention, and customer loyalty. I would ask each of you how much time you dedicate to conducting your engagement research and studies. Do you then ignore the findings? I once did, and it certainly took a while to finally realize.

The strongest brands in the world are aware of this. USAA has historically received higher customer satisfaction and loyalty ratings than its competitors and is currently ranked as one of the most reputable financial companies in the world. USAA's Net Promoter Scores, a key metric to future revenue and referrals, proves their service-focused culture has produced tangible business results: They rank significantly better than industry averages.[2]

Bain & Company research indicates that achieving a mere 5 percent increase in client retention can yield a 25 to 95 percent increase in profitability. Why? Because loyal customers are bigger buyers, retain longer, require less labor, and refer others without compensation.[3]

As we saw in Chapter 6, Costco's model—paying above market wages, promoting from within, and trusting employees to do the right thing—produces some of the lowest turnover rates in retail and membership renewal rates above 90 percent. That stability translates directly to margin, lifetime value, and sustainable growth.

Even Amazon, a company frequently considered to be a technology disruptor, has established its dominance by building a customer-centric, service-oriented culture. Customer experience and customer-centered policies, combined with relentless service simplification, serve as their competitive barrier. Amazon's responsive customer service, transparent service practices, and simple return processes motivate customers to come back and shop more frequently. Transparent communication and proactively resolving issues also generate extraordinary lifetime value.

Customers are not only loyal to Amazon because of ease of shopping, but because caring is integrated into the business model. Efficiency and empathy support each other. Some individuals reading this may have significant frustration with Amazon. For me, Amazon always gets the above right.

According to researchers, most businesses view great customer experiences as essential to their success; however, many are unable to convert these initiatives into measurable financial gains.

Amazon represents what happens when the relationship between the two parties is genuine—the effectiveness of customer-centric business. I find it incredible that I can wake up in the morning, realize I need an item, never leave my house to purchase it, and have it delivered to my door the same day. Although I may have to leave my house to return it, I get the credit right away. Nothing beats that. I am loyal to Amazon, not only because of the convenience, but also because of the simplicity versus price and speed.

Compare that to companies that choose to acquire customers versus earning their loyalty. Each year, that cost continues to climb as these companies spend more and more to replace customers who had no reason to abandon them. Marketing budgets also increase as customer churn grows and margins decline. Loyal customers are more likely to recommend you to friends and family and require less persuasion to make additional purchases. This is the essence of creating economic value.

Returning to the USAA example, USAA has utilized trust as a financial engine and is consistently rated as one of the top in terms of customer satisfaction. The business has consistently demonstrated that it places a priority on its customers—they're more valuable, more committed, and significantly more likely to utilize all aspects of their financial life with the brand (checking, saving, insurance, investments, mortgages).[4] That is pure stickiness they have created. I have worked for one of the largest wireless carriers in the UK, and while we always tried to achieve this, I can tell you it is not an easy thing to accomplish.

Creating a customer-first business provides numerous structural cost advantages as well. Customers are less likely to contact, escalate, and churn when they have confidence in the organization. Every avoidable complaint saves money. Every cancellation that is prevented saves money. Every issue that is successfully resolved reduces other costs, and, over time, these cost savings add up to substantial financial advantages.

Consider the airline industry. Studies on airline customer happiness indicate that customer satisfaction increases substantially when specific service touch-points (even digital service touch-points, such as boarding passes, real-time updates, baggage tracking, etc.) improve. In a business environment with razor-thin profit margins and intense competition, loyalty and repeat business are directly related to the enjoyment experienced by the customer. I recently flew on a major British airline and was unable to track my bags. I traveled across multiple continents and hundreds of hours and spent a majority of the flight asking why? Tracking bags is such a critical digital loyalty enhancer. How is that possible today when compared to many of the smaller airlines that simply get it?

Consider one of those types of airlines like Southwest Airlines. The airline industry is one of the most brutal industries in the world; however, Southwest's emphasis on employee care has provided it decades of success. As rivals go through bankruptcy after bankruptcy, acquisition after acquisition, and bailout after bailout, Southwest's values of humanism, empowerment, and respect have provided its shareholders with sustained profitability, operational efficiency, and customer loyalty. Happy employees not only create happy customers, but they also produce stable revenues.

Companies that have earned the trust of their customers will be forgiven for their mistakes, which will inevitably occur. Unlike their rivals, companies that earn the trust of their customers will be able to forgive themselves for their mistakes and rectify the situation quickly because it demonstrates character and reliability. Conversely, companies that do not cultivate this type of goodwill will find that each failure contributes to the next and will lose their customers faster than almost anything else.

Overlooking return on impact has serious consequences. Companies that under-invest in culture and customer experience will experience slower growth, higher turnover, more service failures, and lower operational stability. Leaders who view customer experience as an expense and not as a strategic asset will ultimately pay for their oversight through reduced revenue, increased costs, and decreased relevance.

Impact management becomes value creation.

This chapter is not a plea for philanthropy. It is a plea for professionalism. The business case for compassion is not discretionary in modern economics. Design for people and the numbers will follow. Trust drives retention. Retention drives revenue. Engagement drives productivity. Experience creates differentiation in commodity markets. Ultimately, all roads lead back to humanity.

Compassion outperforms technology when standing alone. The best organizations illustrate that empathy and efficiency are not mutually exclusive, but rather multipliers. When compassion is developed into the systems of an organization, brand equity explodes. Customers remain longer. Employees are more committed. Reputations are strengthened. Growth is accelerated.

Amazon has created one of the simplest, most customer-friendly retail systems in history. Exceptional lifetime value is created through their proactive customer protection, seamless refunds, and obsessive pursuit of eliminating the pain associated with making purchases. Since it is simple to remain loyal, customers remain loyal. When caring is developed into the system, this is what occurs. When you design for your people, the cumulative positive effect of behaving ethically is that your customers and people will remain loyal to you for longer. End of story. Brand equity explodes.

Not only is doing the right thing the ethical choice, but it is also the financially viable option. The most successful businesses in the world today are those that have incorporated humanity into their core values, and the financial successes of these organizations reflect this. Both the statistics and the testimonials of successful organizations attest to this.

Return on impact is not a soft metric. It is the upstream measure that determines whether your return on investment is sustainable or fragile. The organizations in this chapter—USAA, Costco, Amazon, and Southwest—did not stumble into their financial results. They designed for them by treating trust, culture, and employee dignity as inputs, not outputs.

Here is the diagnostic worth running before Chapter 14: Pick one decision your organization made in the last quarter, like a policy change, a cost reduction, or a technology deployment, and ask what the return on impact was. Not what it saved. What it cost in trust, in employee confidence, in customer belief. If you cannot answer that question, you are measuring half the equation. And that's the half that eventually catches up with you.

CHAPTER 14:

YOUR SERVICE IS YOUR BRAND

For too long, we've pretended there is a clear line between customer service and marketing. There isn't. In today's fast-paced world, service is your brand.

Every time you respond to a customer's email, every time you make a decision about a refund, every time you respond to a customer on social media, every time you solve a customer's problem—these are the moments when you build your brand reputation more effectively than you possibly could through any marketing campaign or even advertising.

In today's world, customers believe others but don't believe you're telling the truth with your advertising copy. That is why reviews are so powerful. Customers trust what they read in a review, what they see on TikTok, or what someone else tells them. And in today's era of instant sharing, one great or terrible customer service experience can reach thousands of potential customers before your marketing team has even drafted a campaign. I have believed for many years, and it is still true today, that word of mouth is the most effective form of mass communication on earth. It is more powerful than any billboard could ever be.

That Lexus moment made my nine years into a now twenty-five-year Lexus customer, and I tell everyone I can about that story, about my burgundy with beige leather, Sheila. Okay, let me make this even more personal for a minute.

Twenty-one years ago, I walked into a Lexus dealership as a young professional to purchase one of my first luxury vehicles. I was in my first senior role. I didn't have a fancy address. I had hope, excitement, and a lot of nervousness about making a significant purchase. The salesperson treated me like I was her most important customer of the day. No pressure. No Ego. Just pure respect.

Years later, when I had a problem with my vehicle, Lexus took care of it, regardless. No debate. No "We're sorry, but..." No "This is our policy..." They simply took ownership and said, "We've got you."

It's probably time for you to hear another loyal Lexus story. You can see my brand fandom come alive again.

When I returned to the U.S. from living and working abroad, shortly after the Lexus experience described above, I purchased a lower model, smaller Lexus. It was the 200 SUV RX range. It was nice; however, I was accustomed to driving the larger 300/400 RX models, so I wasn't sure how I would like it. The outside of the vehicle appeared slightly smaller, but the interior of the vehicle was beautiful, though it certainly had a smaller engine. I purchased it, because a) I was going to be commuting each day by train to San Francisco and therefore would not be driving into the city as much, and b) I thought it would be nice to downsize a bit. After six months, I went back to my local dealership and said, "This is a beautiful vehicle. But I spend all day driving in the turbo mode. It just doesn't have the pick-up I was used to with my last 450." I thought for certain they would look at me and tell me I'm crazy (no comment, readers). Instead, they asked, "What would you like to do instead?" I quickly answered, "Go back to a new 450." We immediately went inside and they traded the vehicle out—no questions asked, no challenge, no hesitation, no judgment.

I left the dealership with a brand-new Lexus RX450. While they did mention that I was the first person to return the 200, they still made it happen and made me feel good about my request and my decision. They respected my knowledge and expertise as a driver. They also did not make me get permission from twenty-seven different levels of management. I was a woman who knew exactly what she wanted, expected, and needed from a drive.

That moment created something that money cannot. I have now been a Lexus customer for more than twenty-plus years. I have recommended the brand to countless friends, colleagues, executives, and family members. I have defended the brand during conversations. I have consistently chosen them again and again.

Great service and great recoveries create great stories. And great stories are marketing gold.

That service moment became my lifetime marketing contract with Lexus. And they have never spent one dollar to advertise on my behalf.

While word-of-mouth has always been effective, the digital platform has dramatically enhanced its effectiveness. Consumers now share online service experiences to express their views and educate others. A single outstanding experience can generate massive amounts of organic exposure.

Here is another beautiful story that has turned me into a loyal fan, without any advertising or marketing. As a pet owner, you are probably familiar with Chewy's gesture when customers share that a pet has passed away—they send them handmade sympathy cards or flowers. Most people don't keep that private. They upload it, post it, cry on camera. The business is tagged, and millions of impressions are created by those posts based solely on caring for a customer. I personally got to experience this when I lost a loving pet.

A Ritz-Carlton hotel had a little fun when a child left behind his stuffed animal. After capturing photos of the giraffe enjoying the resort—getting a massage, looking at the view, driving a golf cart on the beach, the hotel staff sent the stuffed animal back home. The parents posted this story online. The story went global. The Ritz-Carlton didn't create a campaign. They made memories so magical they went viral, and they won more love.

This is exactly what great service does. It makes memories magical, and magical memories become powerful stories—and legends.

Maya Angelou has a famous quote, which I've modified a little as it relates to customers. It still rings true. Customers do not recall what you said, they recall how you made them feel. You can create the most beautiful advertising campaign in the world, but if a customer has to wait on hold for forty minutes the next day, all the value of your marketing equity disappears. Emotional connection, feeling, how you cared for them—positive or negative—is what remains.[5]

Your customer service is your daily brand campaign. Each point of contact is an impression. Each impression is marketing. When we look back on the viral video of a passenger being physically removed from a flight, that negative experience and service encounter erased years of positive marketing, promotion, and polished advertising.

I write this again here. Emotion is the currency of loyalty. Consider this: Your customer care and customer support service teams are actually creating your marketing every single day. They own your brand.

The cumulative effect of providing excellent service and caring is incredible. Perhaps the greatest benefit is that advertising is essentially free to you. Your customers become your media empire when they feel cared for. They post, they share, they tell their friends, they advocate for you, and they recruit for you without you asking them to. You earned it.

The best way to think about marketing today is that businesses with the largest advertising budgets are not the ones that are successful. The businesses that shine are the ones that produce the service moments worth sharing.

Service is your loudest channel, your most believable message, and your highest ROI in our economy built on attention and trust. Advertisements can belong to your marketing team. However, the stories belong to your service team. All day. Every day. Creating a reason for your customers to speak positively about your brand and service today is more than just supporting your customers. It is your most effective, credible, and longest-lasting marketing strategy, and all it requires of you is the courage to care.

CHAPTER 15:

LOYAL CUSTOMERS SPEND MORE AND COMPLAIN LESS

I worked for one of the largest mobile providers in the UK. We were leading in acquisition, as we had sole exclusivity of the iPhone when it came out there. It was incredible. We were winning and winning big. But then, the inevitable happened. Every other mobile provider eventually got the iPhone. And that is when we learned and learned fast that while we can chase the big launches and the growth metrics, retention is where the real wins are—and the money.

I never understand why we don't measure retention statistics and boast about loyalty and for how long. Loyal customers are cheaper to serve, easier to market to, and far more valuable over their lifetime than any first-time buyer. They are also more forgiving when things go wrong. Those companies that acknowledge a customer's loyalty (like Amazon) get it. Amazon Prime members spend nearly double what non-members do. Why? Because the service experience, fast shipping, seamless returns, and consistent communication create trust. That trust reduces friction, and reduced friction drives more purchases. Loyalty is so valuable. We learned that from my Lexus experience. [6]

But why isn't everyone else getting this?

Retention is not always the sexy thing. There are no big splashy press releases or launch day accolades or eye-catching news releases. Yet, it is the foundation of any long-lasting successful company.

If you remember my story about my recent experience with one of the largest specialty apparel companies in the U.S., you will know that my lifetime value could have bought me a brand spanking new Porsche. So, even though things went terribly wrong and I let them know how frustrated I was, I continue to buy from them, stick up for them, and spend money with them. A loyal customer is more than just a repeat customer. They turn into a multiplier. They make more frequent purchases, spend more money each time, try new things without trepidation, and bring friends and family along. The majority of businesses continue to significantly underinvest in retention, despite the fact that it quietly outperforms acquisition over time.

Maybe before we make any big purchases or lifetime moves, we should demand to understand what the customer retention rate is or the lifetime value or repeat purchase rates are.

Starbucks uses a different approach to tell the same narrative. Stars aren't really the focus of their rewards scheme. It has to do with convenience and habit. Starbucks becomes a part of customers' everyday routines thanks to mobile ordering, personalized offers, birthday benefits, and easy payment. Their most devoted customers come in many times a week. One of the most enduring customer bases in retail and billions of dollars in recurring revenue have resulted from that retention-focused ecosystem.

Apple doesn't even have a formal loyalty program—it's their service experience. Products are smoothly integrated. Support is reliable. Fear and friction are eliminated by Genius bars and everyday enhancements. Because of their trust, not points, customers will wait overnight for their next release, which leads to exceptional retention and a high lifetime value. I recently upgraded my cell phone, and I have told many people how much I love my phone and how easy the transfer experience was. That was what used to make me slow to adopt a mobile change, more than most. Now, I am not afraid of it and even more loyal to their phones.

This is what differentiates gaining loyalty from purchasing it.

Too many businesses think that loyalty can be bought with gimmicks, points, or discounts. However, no rewards scheme in the world can make up for being difficult to get in touch with, sluggish to react, or negligent when issues come up. Consistency and concern are the foundations of loyalty. Trust is built on care. Connection is based on trust. And what keeps customers coming back is connection.

Loyal consumers behave differently in addition to staying longer. Because they don't want to end the relationship, they complain less. They expect you to put things right, therefore they don't escalate as much. Instead of attacking, they offer comments with the goal of improving. Since you have already earned their goodwill, they presume it. If those customers do complain, they are not truly loyal.

Service expenses decrease when there is confidence. Escalations get smaller and simpler to market. Growth stabilizes. The company relaxes more.

Compare it to a large, big-box retailer that introduced an ostentatious loyalty program and promised huge discounts and exclusive benefits. The first year saw an explosion in enrollment. However, the underlying service experience was flawed, with lengthy checkout lines, perplexed staff, erratic return policies, and inadequate communication. Retention fell apart in less than a year. The experience was more important than the discounts. Loyalty is not created by loyalty schemes but by experiences.

It is not enough to measure customer loyalty based solely on metrics like CSAT scores or NPS data. Companies that understand what drives customer loyalty, and therefore ultimately increase revenue, track the true signs of customer loyalty: the number of years a client has stayed, how much overtime, how often they come back, how many referrals they send, and how actively engaged they are. These are the monetary footprints of trust.

Even if it seems boring on an investor presentation deck, retention is still the greatest investment a business can make. Caring for customers is how trust is built. And trust builds loyalty. So, loyalty creates more sales and therefore, it is the simplest business growth strategy available today.

But companies will suffer a great cost for ignoring this truth. Executives know about cost; however, most executives grossly underestimate the cost of apathy.

Apathy does not hurt a business with one big, public failure. Instead, it slowly seeps out. It begins to cost money with each lost customer. Each customer has also stolen future money (the next year's worth of purchases), lost referrals (recommendations), lost upgrades (improvements), lost renewals, and lost advocacy. The worst part is that lost customers also steal the money you spent marketing to get them back, and that marketing money is now wasted trying to replace what should never have been lost.

Service inflation is the second cost of apathy

Disengaged customers complain more, escalate faster, and need more time and resources from your employees. Your employees spend more time on calls. Your managers spend more time fighting fires. Your legal team and public relations team show up in conversations that used to be easily resolved with a simple apology and solution. What would have cost $20 to resolve now costs $2,000.

Employee burnout is the third cost of apathy. Angry customers become employees' problems. Employees burn out when they are not empowered, trusted, or supported. Employee burnout causes employee turnover. Employee turnover is expensive. Employee turnover requires frequent rehiring. Frequent rehiring makes service inconsistent. Inconsistent service increases customer frustration. The cycle continues, and the business suffers losses due to decreased productivity, rehiring costs, training costs, and the loss of institutional knowledge.

Brand deterioration is the fourth cost of apathy. Marketing can hide cracks for a while. But eventually, reality sets in. Customers write more negative reviews. They speak louder on social media. And they lose faith in the brand. Once customers lose faith in a brand, no amount of advertising dollars can build faith in the brand as fast as it was lost.

Momentum loss is the final, and most deadly, cost of apathy. Businesses that don't care about their customers and employees stop improving. They stop paying attention. They stop noticing when their customers and employees are subtly disengaging. Their growth stops. Their innovation stops. Leadership is not about producing results anymore; it is about explaining the results while competitors quietly steal their customers.

I remained loyal to the retailer after things went wrong because I truly believe they care. I spoke directly to the SVP of the retailer, and they fixed my problem immediately and did the right thing. That is why I remained loyal.

That explains why indifference is so costly and loyalty is so powerful. Loyal customers are less likely to leave. They give you time to fix mistakes. They protect your brand from attacks. While other customers leave, loyal customers stay.

Caring is not a soft concept. It is a tangible, competitive advantage for businesses. Businesses that fail to learn that lesson will always pay the price: in culture, in churn, in brand deterioration, and in momentum loss.

Loyal customers are more profitable. Loyal customers complain less. Loyal customers are cheaper to serve. Loyal customers create consistent and sustainable revenue growth.

Retention is not sexy, but it is the smartest money you'll ever spend. Loyalty is where the real value is created.

PART V–
BUILDING A COMPANY THAT LASTS

Businesses are not designed to last. Most are built for the next quarter, not the next decade. Part V is about changing that.

CHAPTER 16:

BUILD FOR THE LONG HAUL

Outstanding service cannot be a campaign or program; it has to be a core philosophy of your business—one that shows your concern for the well-being of both your customers and your employees, and a commitment to using your systems, technology, policies, and procedures to provide the best possible experience for every one of those people.

It is easy to win in the short run (i.e., quickly cut costs, automate processes aggressively, raise short-run sales and volume); however, there are consequences when your business seeks to be efficient at all costs. Your company may be able to operate cheaply for some time, and while you might think you're being effective, you will lose your ability to innovate, your business will continue to decline in relevance, and your business will eventually be gone.

On the other hand, if your business chooses to focus on becoming "built-to-last"—a business that remains relevant regardless of changes in the marketplace; retains its customers; provides a level of value that encourages employees to believe in the company for years to come (and long after the current leadership has left); increases long-term value rather than just short-term statistics—then you will be successful, and your business will endure for years to come.

We now understand that caring is an invaluable strategy. Service is the operating system of your business, and what you need to do is stop just reacting and begin to build and endure for the future.

A customer experience is not a quarterly initiative. It is not a turnaround project. It is not a digital transformation program with a ribbon-cutting ceremony and a press release. It is not something you "fix" and then forget. A customer experience is your operating system. It is the base on which everything else is built. Companies that remain successful for decades, not just for reporting periods, build the customer experience into their structure from the beginning and continue to build for the long haul.

Too many companies do not want to think long-term. As a result of needing to be fast, many companies sacrifice the long-term benefits of durability, trade trust for short-term profits, and sacrifice long-term growth for the short-term optics of headlines every quarter. Rather than focusing on creating a corporate culture, many companies focus on optimizing for the latest headline.

Many companies will mortgage the future in order to capture this month's statistics—and when the future arrives and relevance is gone, and the culture is gone and allegiance is gone, companies will claim to be shocked.

Companies that have a long term vision from the very beginning have a completely different mindset. No matter how much pressure they receive, they make decisions based on what they will and will not compromise. They invest in developing their leadership before it becomes a necessity. Institutional knowledge is considered important above continuing restructuring. They create incentives for their employees to act as stewards rather than solely focusing on the immediate gains. They realize that sustainable progress is designed, implemented, and grows slowly, deliberately, and with no intention of ever stopping.

Building for the long-term requires discipline and a tremendous amount of patience. It takes time. It requires leaders who are willing to invest long-term in relationships, systems, and people that may not produce immediate results. Organizations do not improve themselves when they continually patch old employees, add new technology to ancient systems, and cover broken rules with band-aids. They are merely delaying their downfall. Six months later, they return to their original condition; however, their volume, cost, and fragility have greatly increased.

Businesses often treat the customer experience like a leaky faucet. They fix the leak when it happens. They respond to complaints when they go viral. They buy new software when churn increases. However, the underlying problems remain. The culture remains strained. The incentives remain misaligned. The small cracks eventually turn into major structural problems. No amount of marketing can cover those problems.

I have lost count of how many organizations I have worked for that wanted to constantly change and place blame on the systems we just implemented. Rather than repairing the fundamental issues, maintaining the employees who have endured these changes, and providing a solid base of operations, they pursue short-term solutions. Working for numerous organizations over the years has resulted in my number one heartache on the client side.

Legitimate long-term builders work much harder. The foundation is repaired. Instead of constantly searching for new customers, they develop the technological infrastructure, training programs, leadership methods, and organizational policies that adapt with their customers. They provide long-term care, not project-based care. Rather than replace employees, they utilize technology to augment them. They create habits of relevance rather than singular events of relevance.

Additionally, they institutionalize the process of learning. Mistakes are viewed as data, not a cause of embarrassment. They create feedback loops that identify potential problems prior to them turning into actual problems. They encourage curiosity. They reward and value it. Rather than becoming increasingly defensive, they create communities that evolve with knowledge over time. Constant learning is a strong indicator of a long-lived organization.

One of the greatest examples of a business that thinks this way is Macy's. For nearly a hundred years, Macy's has survived the factors that destroyed traditional retail, including the decline of malls, the rise of e-commerce, changing consumer expectations, economic downturns, activist shareholders, and a global pandemic. Many of its competitors went out of business. Macy's adapted. They are currently navigating one of the most significant restructurings in their history—closing approximately 150 underperforming locations to invest more deeply in the stores that remain. That is itself an act of long-term thinking. Rather than cling to a model that no longer fits, they are making difficult decisions to preserve the brand's future. I am confident they have not gotten everything right. But they are still in business, still evolving, and still choosing durability over drama.

Macy's did not sustain itself because it was the first, the cheapest, or the biggest. They are still here because they evolved without sacrificing their essence. They preserved the emotional impact of in-person interactions while investing in retail renovations. They used both physical and digital mediums to create a seamless customer experience. While not disrupting loyal customers, they improved their own branded products and partnerships. They upgraded their fulfillment, commercialization, and loyalty processes while honoring the traditions of the brand, such as the Thanksgiving Day Parade.

Most importantly, they preserved the emotional connection with their customers. Platforms do not connect people. People fall in love with the emotions a brand elicits in them. Rather than replacing humanity, Macy's innovated on top of it. Building for the long-term looks like that balance.

Nordstrom shares the same mindset. For decades, Nordstrom has emphasized empowering employees, trusting them to use their discretion, and placing trust in employees' judgment above rigid efficiency. Their well-known service culture is not accidental. It is the direct result of executive leaders choosing to prioritize long-term loyalty over maximizing profit margins in the short-term, prioritizing relationships over regulatory compliance, and prioritizing judgment over scripted responses. Multiple generations of customers have developed resilience due to that commitment.

Garry Ridge, the former CEO and present Chairman Emeritus at WD-40, exemplifies long-term planning more effectively than any other executive. Under Ridge's direction, WD-40 established capabilities instead of simply pursuing expansion. Ridge spent years establishing a culture of accountability, psychological safety, and learning. He reframed mistakes as "learning opportunities." He made a large-scale investment in coaching leaders. He established trust as part of the normal course of business, as opposed to an annual ritual. Instead of merely posting values, Ridge ensured the values of the company were operationalized.

Caring was not the ambition at WD-40. Caring was designed. Instead of preparing to lead, leaders at WD-40 learned to serve. Employees were encouraged to test new ideas responsibly. Instead of fearing criticism, employees were encouraged to share feedback. Employee performance evaluation systems fostered teamwork, as opposed to competition. That ultimately created an organization that was adaptable without becoming disorganized. The strength of the culture was established before the financial success of the company, not the reverse. If you wish to read a book on this topic, please see his *Any Dumbass Can Do It.*

Costco built for trust, as opposed to extracting profit from customers. Costco has developed one of the most resilient retail models in the world by making a choice that most retailers fear to make: It consistently prefers fairness over short-term removal. Costco has developed one of the most robust retail platforms in the world. Costco limits price markups and pays employees fairly and promotes internally. Costco does not pressure vendors for minimal profits. Critics argued for years that this method would reduce profits. On the contrary, it generated remarkable stability. Low employee turnover results in higher levels of employee productivity. Employees who have had more experience provide better service. Customers renew their membership at historically high rates because of their faith in the brand. Costco illustrates that loyalty becomes structural rather than situational when customers are treated as long-term partners, not transitory business associates. Based on their model, trust is not a by-product. It is the business plan.

Patagonia is another great example of a company that took values and turned them into a lasting benefit. Patagonia has demonstrated that one of the strongest competitive advantages in business is purpose, when applied consistently. Rather than constant consumption, Patagonia encourages repair, reuse, and responsible purchases. It invests in environmental preservation on a massive scale. It takes public stances on issues that other companies shy away from. Patagonia operates in accordance with its principles regardless of whether the implementation appears risky. This is not a marketing show. This is an operational discipline. Customers have faith in the company because of its consistent adherence to value. Patagonia can command premium pricing. It has created a zealous level of customer loyalty—it's multi-generational.

Patagonia illustrates how legitimate and ingrained values do not hinder but stabilize and increase momentum.

Toyota builds for longevity by using learning. Toyota's resilience is a direct result of its unwavering commitment to education and development. Kaizen, Toyota's well-known principle of continual improvement, is neither a program nor a slogan. It is a discipline that is utilized on a regular basis. Every employee has the authority to identify problems, submit improvements, and prevent defects from reaching customers. There is no provision for shortcuts. Quality is guaranteed. Over time, employees gain knowledge and maintain it. This provides Toyota with the capacity to absorb shocks, adapt to changing market conditions, and maintain customer confidence across generations. Toyota invests in employee capabilities, while many competitors focus on the next cycle of rapid expansion or reducing costs. Toyota develops organizations that become stronger, smarter, and more resilient over time. Building for the long-term looks like that.

Long-term organizations build humans in this way. They embed learning into their systems, protect their cultures, treat customers as long-term partners, and prioritize the development of their employees in succession planning, performance evaluations, hiring, and promotions. They do not rely on heroic leaders to carry the culture. They create the systems that sustain it. The values of the organization do not change with the leadership of the organization. The organization adapts to the changing market conditions. The trust of the organization remains during times of crisis.

Organizations that thrive do not deal with symptoms. They rebuild the causes of the symptoms. Starbucks did not simply hire more employees when they experienced long lines and conflict. They redesigned the experience of ordering via mobile platforms. Toyota did not contractually delegate responsibility while it was under pressure to improve quality. They embedded ongoing development into the routine tasks of their employees. These organizations did not react. They redesigned.

These organizations also understand that what currently satisfies customers will soon become the norm. The value of excellence decreases.

Customer expectations increase. Organizations must perpetually reinvent themselves to ensure relevance for the future. Organization leaders must anticipate customer demands before they become complaints, establish flexible policies versus rigid policies, develop the skills required by employees for the future versus the past, and create technology that enhances, rather than hinders, existing practices.

Amazon represents the mindset described above. Regardless of your view of Amazon, the long-term strategy of the company is undeniable. The customer experience is never "complete." Voice commerce, one-click ordering, predictive logistics, and frictionless returns represent an ongoing commitment to convenience and confidence. Compare that to Kodak, a once-dominant brand that failed to adapt to the digital era. They were not saved by loyalty alone. They were saved by their relevance.

This should serve as a wake-up call for everyone. Your customer experience strategy was created for this quarter if you are preparing for a downturn. You are damaging your future if your incentives focus on short-term gains, rather than long-term potential. The definition of building for the long-term is creating systems, rules, training, leadership behaviors, and performance metrics with the customers you will serve five years from now in mind, not just the ones you billed last quarter.

In some organizations, we refer to this behavior as playing "Whac-A-Mole," a game where you either solve a problem or a mole pops up, only for another one to pop up elsewhere. Short-sighted solutions never solve long-term problems.

I personally saw this in the hospitality industry. We built for audits, checklists, and quick wins far too often. It all seemed strategic. It was almost never architectural. Even though we talked about how customer needs were changing, we rarely built with those changes in mind. Sustainable service requires more than just talking about it. Structural commitment is required. To delay their downfall, only more outdated approaches and longer timeframes for solving for the customer of the future will occur by being risk-adverse and constantly seeking the perfect system or platform or approach.

What ties together companies like Macy's, Nordstrom, WD-40, Costco, Patagonia, and Toyota is not their industry, size, or business model—it's their mindset. Each of these organizations intentionally chose to build for durability, not drama; for trust, not headlines; and for capability, not convenience. Rather than pursuing every trend, sacrificing values for quarterly optics, and extracting short-term gains at the expense of long-term relationships, each of these organizations committed to investing in their employees, embedding learning into their systems, protecting their cultures, and treating their customers as long-term partners, not temporary transactions. None of these organizations stumbled into longevity. They engineered it. And the lessons they offer are not abstract. They are practical, repeatable, and available to any leader willing to think beyond the next reporting cycle.

There are some big differences between long-term builders and short-term builders in terms of how they select new technologies, automation tools, and their technology stack. They do not fix things. Long-term builders build with scalability in mind. They plan for the future. They make an investment in flexibility. Rather than responding, they take the lead.

Longevity is not a result of good luck or good fortune. It's a deliberate design. Change your resume if your company doesn't care enough to invest in the future. Because businesses don't get second chances if they don't play the long game. They will just get replaced.

CHAPTER 17:

YOU DON'T HAVE TO DO IT ALONE

Here's a powerful truth most leaders are reluctant to admit: You cannot always do everything yourself—at least not very well—sustainably or at scale. Your teams are overwhelmed; your budget is limited; your customers expect Amazon-level service delivery; your competitors are partnering, automating, and using AI to become more agile and efficient. Trying to bear the entire burden of doing everything yourself is no longer a badge of honor. It is instead a liability. But if you do it the right way, you can scale smarter, use business process outsourcers (BPOs), and advance with good automation and AI.

Companies that succeed and thrive in the next ten years will not be those that attempt to control every function and activity end-to-end. They will be the ones that demonstrate an awareness of what they are good at; an appreciation for areas in which they require help; and an intentionality around how they utilize partners and technology to augment, rather than displace, the human experience. Asking for help does not diminish your culture. Refusing to acknowledge your need does.

You won't lose your soul by getting or asking for help. You will by trying to constantly protect your ego.

BPOs are NOT the Enemy

Outsourcing has been a bad word for far too long. In boardrooms, leaders whisper, "We might need to outsource," as if they have failed in some way. Others view outsourcing only as a crude method of cost reduction, reducing services to their most basic or worst forms and then being shocked when customer loyalty wanes.

I have worked with more than twenty-five BPOs in my professional career and then joined one. I understand that BPOs should be viewed as a partner and not merely a cost-cutting measure. When BPOs are used appropriately, they can bring about the necessary level of consistency, reach, and resiliency to provide you with a competitive advantage in today's business environment.

In fact, world-class companies view BPOs in this manner and utilize them to create and grow their businesses. They view BPOs as an opportunity to enhance their innovation and support their business goals rather than simply looking for ways to cut costs. Some of the other benefits include expanding your multilingual support and managing demand spikes by quickly scaling your workforce. It also helps maintain business continuity during crises, such as natural disasters, and enables you to automate processes and implement AI in areas that may otherwise be too time-consuming or costly.

I have witnessed numerous travel industry companies remain stable throughout and after Covid because of the existence of a global BPO network, which enabled them to support multiple languages across various geographies, manage rebooking, and cancel and refund customers across continents and languages. This was something that no single internal operation could have managed alone. I utilized a BPO to increase sales in the home services industry during the pandemic when we were 90 percent above our call forecasts and needed to quickly hire new staff to handle the increased volume of calls.

A BPO is not just a way to shave pennies off your customer service budget. A BPO provides you with scale, resiliency, and expertise in areas of your business that you may not be capable of doing on your own. Whether it's due to a budget, bureaucracy, AI-driven steering committees, peak season, product launches, or social media, a BPO can help provide you with the capability to meet your business objectives.

A BPO can also provide you with the ability to rapidly scale your workforce during periods of high demand and viral moments; access to specialized industry.talent; multilingual global coverage twenty-four hours a day, seven days a week, 365 days a year; and operational rigor through strict quality assurance (QA), service level agreements (SLA), and training.

Your in-house team may become overwhelmed during high-demand situations. A BPO can deploy hundreds of trained agents in a matter of weeks. The things you may aspire to deliver internally are "non-negotiable" to a good BPO partner. The best BPOs have helped me save money and enabled me to handle challenges I once thought I couldn't meet—delivering faster and with better results.

A BPO is designed to attract and retain specialized talent. Healthcare? Travel? Finance? The right BPO will have employees who are trained in your specific area of business. You won't have to reinvent the wheel. A BPO will allow you to have day and night coverage across time zones, languages, and geography—because your customers don't stop sleeping.

Again, having worked with over twenty-five BPOs across different industries and geographies—and currently employed by one—I can confidently say that many of the largest operational accomplishments I have made would have been impossible without them. I have experienced positive results from my partnerships in regard to accent neutralization, multilingual support, regulatory compliance, revenue growth, and reducing complaints. Working with a BPO is not about losing control of your business. It is about obtaining capabilities you don't possess and speed you can't match.

Automation and AI Are Not the Enemy Either

AI can never replace your humanity. That's a phrase I plan to adopt and use throughout the industry going forward. It is here to assist you. Technology and humanity do not have to be at odds; they should complement each other, and people must continue to care, regardless of how much technology you use. This is the most important thing I want to communicate with you today. Never replace but always support and absolutely complement your employees.

When an organization deploys automation to avoid customers, this is what happens to their loyalty—it drops. When an organization uses it to reduce friction, empower teams, eliminate complexity, and make things simpler, the organization builds trust in its customers.

Smart organizations use AI to route and triage, so customers don't bounce back and forth across multiple departments. They use it for agent-assisted tools so their representatives don't have to juggle seven systems, a binder of scripts, and twenty-seven different screens at one time. The smartest organizations use automation to proactively notify their customers of routine updates: "Your flight is delayed. Here are your options." AI is a great tool, as long as you're using it to enhance how you:

- Route customers to the correct area the first time
- Provide real-time assistance to agents
- Automate proactive notification
- Automatically resolve simple issues immediately
- Reduce system overloads and manual work

My local air-conditioning company is also using it correctly, providing automated scheduling and booking for service. My waste collection company texts me when they're running behind schedule and then follows up to ask me to confirm a new schedule.

Starbucks did not fix morning rushes by adding more staff. They developed mobile order and pay. This was a game-changer for convenience, allowed for larger ticket sizes, and strengthened their customer loyalty. Chewy automates reorders perfectly, American Express utilizes predictive IVR to route calls to the right specialist quickly, eliminating the need for their customers to scream "representative" into the phone. Self-service done correctly provides large customer wins. AI-powered FAQs, chatbots (when done well), and customer knowledge bases can address the majority of your most common issues. In addition, you'll win with effective call-listening automation if you apply the data effectively.

Self-service done correctly can answer 60 to 70 percent of basic customer requests with minimal frustration. Cognitive overload and handle time can both be improved with agent-assist tools. Predictive systems can prevent issues from occurring.[7]

Good technology creates trust; bad technology destroys it. Intelligent technology creates space for compassion. It allows humans to connect, listen, and care. Consider automation and AI as force multipliers, not replacements.

The Future is BPO and AI together

If you need help, the future of service/sales won't be BPO vs. AI; it may very well be BPO and AI combined. This partnership can assist you in scaling service or sales without sacrificing your "soul." As forward-thinking BPOs begin to realize their role in the future is no longer as employment agencies but as innovation partners, they will begin to integrate AI into workflows and utilize real-time coaching to create opportunities for faster and improved agent training due to AI simulation and provide insights at scale.

You may ask yourself, "Well, isn't she just saying this because she has been paid to?" No way—not true. I have used BPOs throughout my career because they could deliver what I needed faster and cheaper than my own company's in-house technology teams and long list of approvals processes. Additionally, because I care about my employees/agents, I certainly utilized the resources provided by my BPO partners effectively.

I would not have been able to accomplish much using our internal teams. However, I was able to use my BPO partner to complete those tasks much faster than any of our internal teams could have. It wasn't because I didn't have excellent teams or because I didn't have an excellent group of tech people. It was often because of the massive approval processes that exist in many large companies, or because we simply couldn't afford it this year... and so on. Does this sound familiar?

My BPO partners were able to roll out these solutions faster, with fewer bureaucratic approval processes. Here are some examples of what an integrated model, such as the one described above, could potentially allow or enable you to do:

- **AI + BPO for scale:** One BPO partner rolled out AI-assisted chat for a retail giant. Seventy percent of inquiries were resolved without human escalation, while the remaining 30 percent got to agents faster with full context. The result? Lower costs *and* higher customer satisfaction.
- **Real-time quality assurance:** I spent two years at one company making the business case for this and multiple proposals, demos, etc. Instead of reviewing 5 percent of calls manually, AI can score

100 percent of interactions instantly. That means better coaching, less subjectivity, and faster improvement.

- **Predictive demand modeling:** Smarter staffing prevents longer wait times and service breakdowns.
- **Culture at scale:** With AI monitoring tone, empathy, and consistency, you can ensure your brand voice is reflected across millions of interactions.

This is how you scale without losing your soul.

Excuses Kill Growth

Customer service will be about experience, not price.

I would have spent those extra dollars on a new handbag by now for all the times I've heard other leaders state that they will lose control of their business if they outsource. With the right partners and tools, you'll find that outsourcing will actually give you far better visibility and accountability from your vendors than many of the fragile internal systems that you currently have in place.

What matters most to your customers is how quickly they can resolve their issues and how they feel at the end of their phone call, not who answered the phone.

A leader who fears using an AI tool for fear of being perceived as cold or robot-like is simply creating the experience they want to avoid if they choose to reduce costs and staff levels, rather than focus on how AI can help the customer.

To scale smarter, leadership must be honest. This occurs when you audit the areas where your customers are waiting, where your employees are burning out, where your money is leaking, and where the complexity associated with your organization is killing both customer loyalty and employee commitment.

PUTTING THESE PRINCIPLES INTO ACTION:

- Identify clearly what aspects of your organization require the level of brand intimacy that only in-house solutions can provide, and identify those items that can be scaled more rapidly through partnering with external organizations.

- Partner especially when there are opportunities for greater scale and expertise.
- Test automation before expanding.
- Take the pressure off your front-line team members.
- When selecting a partner, consider their value system, not just their pricing models.
- Make sure that cultural compatibility is non-negotiable.

Great BPOs Care Deeply About Your Customers and Your Business—Competing on Price

Here is the truth: You can't beat your competition on price. You can't beat your competition on price. So true, I had to write it twice. However, you can beat your competition in experience.

The only way to achieve that is to stop trying to do everything yourself. Utilize the best partner available when you need scale. Automate when you need speed. Utilize AI when you need intelligence while you continue to protect your company culture as much as possible. Data, insight, and analytics are king and will help you know what you need to do.

Your customers don't care how you provide a good or bad experience. Your customers care that you provide excellent experiences in the way they wish to receive it and that you follow the platinum rule. Your customers care that you provide it, and that you do it well—quickly, fairly, and with true empathy for their needs. I am confident the companies that will lead in the future are those that may not be the lowest cost providers.

They're going to be the ones that provide seamless, scalable, and human service. That can only happen when you stop trying to do everything yourself. That is the future of customer service. And if you don't develop it, rest assured that your competitors will. That is how businesses endure. The future of providing customer service belongs to those who have the courage to seek help and the wisdom to utilize it properly.

It is the leaders who have the courage to seek assistance and the discernment to make effective use of it who will be leading in the future.

PART VI–
REIGNITE CARING AND FAN THE FLAME

Now the question is not whether caring works. The evidence is in every chapter behind you. The only question left is whether you will have the courage to act on it.

CHAPTER 18:

THE CALL TO ACTION TO CARE AGAIN

You have looked into the many systems, policies, technological innovations, and leadership choices that either enhance and protect human dignity or sometimes imperceptibly reduce and ultimately destroy it. You have also learned that growth without a sense of humanity is meaningless; efficiency without empathy creates isolation; and scale without trust creates vulnerability and instability. Each of these chapters has shown that there is one irrefutable truth: Compassion is not an outcome of good intentions, but a discipline; a design choice; a process that leaders must proactively create, defend, and revive daily. Leaders and organizations must deliberately establish, protect, and nurture humanity at the expense of efficiency.

Rekindling caring is not sentimentalism or a nostalgic attempt to restore compassion. It requires boldness. It means identifying individuals in environments where they are encouraged to cut corners. It means knowing when technology and/or AI will help humanity and when it will begin to replace it. The most successful leaders know when technology and AI remove emotion and friction. They know when efficiency enhances an experience and when it slightly diminishes it. Rather than confusion, distance, or indifference, they use innovation to promote clarity, speed, and confidence. Rather than allowing systems to justify their existence, they ensure that every system, procedure, and tool promotes connection, not disconnection.

But if compassion is not intentionally preserved, it will fade like any flame. A dramatic event does not collapse a culture. It deteriorates incrementally through numerous small ones. It erodes through the unheeded complaint, the rushed conversation, the new policy designed to prevent customers from resolving their own issues. It deteriorates through the employee who remains silent and through the leader who, "this one time," places the value of expediency above morality. All these occurrences accumulate over time until an organization loses its distinguishing characteristics. Revitalizing the flame is to be aware of those moments, to intervene, to return to established standards, to choose courage over comfort, and to preserve humanity regardless of the presence of others.

The subject of legacy is the focus of this final section—about the difference between building something meaningful and running an organization, about deciding to inspire trust or simply measure performance. Organizations that disrupt entire industries are not identified by their quarterly numbers, dashboards, or algorithms. They are recognized for their ability to make people feel good, how they treat people when they are under pressure, and how much they protected the dignity of humans in an increasingly mechanized world.

At this point, you should no longer question whether caring works. That data is contained within the chapters leading up to this point. What is left to determine is whether you will have the courage to lead this caring revolution forward.

Throughout this book, we have faced what works and what subtly undermines organizations from the inside out. We have examined culture, policy, leadership, technology, recovery, loyalty, and trust. Excuses have been dismantled. Comfortable myths have been destroyed. We have repeatedly demonstrated that compassion is neither emotional, gentle, or naïve. It's a calculated move. It can be measured. It makes money. It is also imperative. The only question that remains is, "What are you going to do with all this?" Because it is worse to know something and ignore it than to be ignorant of it.

The future is arriving faster than most organizations and leaders are willing to admit. AI will become the front door to nearly every customer interaction. The foundation of service delivery will be automation. Personalization will no longer be as impressive as it is today; instead, it will be expected. Reactive firefighting will be replaced by proactive service. In a few years, most customer life cycles will start with AI, and customers will want those technologies to identify, comprehend, and assist them immediately. "Sorry, I didn't understand that" won't be accepted any more. Yes, I have also yelled at too many IVRs. Disconnected, emotionless, and poorly designed systems will not be acceptable and will be penalized perpetually.

At the same time, human employees will continue to be more valuable, not less. Routine work will eventually become automated, and what remains will be complex, emotionally charged, and high-stakes, difficult work. In real time, your frontline teams will become brand advocates, issue solvers, and trust builders. No matter how sophisticated the technology develops, businesses that don't invest in empathy, judgment, and communication will fail. Machines are not the only ones that will shape the future. It belongs to leaders who understand how to bring humanity and intelligence together.

The most thriving businesses will function in hybrid environments. BPOs will continue to evolve into partners of innovation. They will combine process design, analytics, and AI. They will jointly own the results. They will not only absorb volume; they will also assist in determining strategy.

The concept of hyper-personalization will become commonplace. Customers will expect brands to anticipate their needs before they articulate them. In fact, many customers like me already expect this. Generic experiences will be disrespectful. Proactive service will be the norm. Culture will continue to be the differentiator in all of this. Your tools can be imitated by your rivals—your character cannot.

I can't express this enough. Humans versus machines is not the future of customer experience. Humans will be with machines, using machines. Yes, happy to take on Elon Musk over this. Ready for the good fight. Let's go.

We soon won't talk about whether it is in-house or outsourced. Instead, we will strike the ideal balance between the two. BPOs provide you with scale. You get speed from AI. You get your soul from your culture. You will lead and win if you get all three right. If you make a mistake, you will fall behind. Technology without humanity becomes cold. Humanity without capability becomes fragile. Great companies will master this. This isn't a motivational speech. It is a cautionary tale.

The marketplace is harsh. Your customers are impatient and have a plethora of options. While you "figure it out," they won't be required to remain faithful to you. You are teaching your customers to leave every time you disregard a complaint, cover up with a policy, put off repairing a flawed procedure, or prioritize convenience over customer service. They're listening too. Theory has never been the focus of this book. Action has always been the plan. Do you desire loyalty? Earn it. You want to make money? Improve your service. Do you want to be relevant? Show greater concern. Care MORE! A few negative reviews are hardly the price of doing nothing.

Start here, right now, if you're serious about caring again:

- Walk your own customer story and experience path. From your first click to your last support, experience your business just as your customers do. Trust me, it will irritate them more than it irritates you.
- Fix one ignorant rule this week. Not in the upcoming quarter. Not following a meeting of the committee. This week. Show that you're serious. Prove it.
- Listen while hanging with your team. Find out what's broken. Pay attention without being defensive. Then take action based on what you hear.
- Make a single change noticeable. Show progress to your clients. Trust is developed more quickly by transparency than by perfection.
- Utilize technology for a reason. Automation and AI should be used to reduce friction rather than accountability. Instead of replacing empathy, let systems enhance it.

When little things are done repeatedly, they develop culture. Culture becomes an advantage when it is maintained over time.

It feels secure to wait. It's not. Others are building while you debate. Others are getting better while you analyze. Others are gaining trust while you hesitate. You can't outspend your rivals. However, you can outsmart them. In business, that is still the most available edge.

Get your act together. This time, for real. Not in a memo or in a slide deck. Not in a rebranding but in your leadership style. Do it in the manner that you choose, in the way you handle others. Let's clean house.

Eliminate bureaucracy. Encourage judgment. Give bravery a reward. Preserve dignity. Use technology to help, not to conceal. Additionally, treat each and every customer like they matter, because they do.

I have been in the room when companies decided to stop caring. Usually not in those words, and usually not all at once. Rather, it happens in the small decisions. The policy that never gets revisited. The complaint that gets closed instead of learned from. The employee who raises their hand and stops getting called on. I have also been in the room when companies decided to start caring again. It can be a small first step, like one leader asking one honest question: What are we actually doing to the people who work for us and the customers who chose us?

That question is yours now.

THE BECKERMAN CARING CHALLENGE™

A TWELVE-MONTH COMMITMENT TO RESTORE AND REIGNITE HUMANITY

This book is meant to be "lived" and "activated," rather than just something that can be "read" and then set on a shelf.

The Beckerman Caring Challenge™ is your personal call to action—an invitation to turn the insights from this book into meaningful action. It's your line in the sand and a public declaration that caring is no longer optional, but your new standard.

That is the moment when insights become discipline, intention becomes execution, and your values move beyond words, slides, or pictures on a wall to show up clearly in your everyday actions. Over the next four quarters, you will lead differently. You will make decisions based on humanity, not solely on convenience and/or efficiency. You will replace habits that destroy trust with systems, policies, and practices that create loyalty. You will build cultures in which people are seen, valued, respected, and feel safe to provide their best.

You will develop the capacity to care for others, lead with courage, and obtain commercial success without sacrificing your soul.

The Beckerman Caring Challenge is a simple way to turn your convictions into actions and demonstrate your values through observable leadership behaviors. Leadership is not about choosing between your performance and your people; it is about developing mastery over both. The Beckerman Caring Challenge will push you past your limits, reveal your blind spots, hold you accountable, and ultimately change you.

This is not about striving for perfection; it is about continuing to move toward purpose... toward growth with integrity and success with humanity.

Welcome to your journey forward. Welcome to a new standard. Welcome to the Beckerman Caring Challenge.

QUARTER 1:
REAWAKEN—SEE WHAT IS REALLY HAPPENING

Focus: Awareness, Honesty, and Accountability

To make any changes, you must understand what needs to be adjusted. Most leaders are working through filters—dashboards, reports, assumptions, and usual ways of doing things—that distort both reality and actual experience of the customer and employees. The theme of this quarter is removing these types of filters so you can get a clear and honest picture of how your customers and employees are experiencing your organization.

THOUGHT-PROVOKING QUESTIONS

- When was the last time I experienced all aspects of my business from beginning to end as a customer would?
- Where do we regularly frustrate people and think that it is acceptable?
- Where do our policies protect ourselves, but not our customers?
- Who on my team feels they are being heard and why?
- Would I trust us if I were my customer?

THE 90-DAY CHALLENGE

- Experience your entire customer experience path both digitally and in person at least twice.
- Attend at least twenty frontline interactions and avoid intervening.
- Remove one policy that is causing distrust.
- Host "Tell Me the Truth" meetings with employees every month.
- Publish at least one process improvement and track the results.
- Evaluate whether AI, automation, or self-service is stopping customers from completing a task or in some way improving and/or enhancing their experience with you.
- Respond personally to at least ten unresolved customer complaints.

COMMITMENT TO CARING

- I will find out the truth, not just what is comfortable or easy for me to believe.
- I will stop hiding behind reports and begin listening to people.
- I will address the damage done by brokenness without becoming defensive.
- I will hold other colleagues and team members accountable for their actions and not defend them with excuses.
- In our organization, caring means that we will always__________

QUARTER 2:
REBUILD–FIX WHAT IS BROKEN

Focus: Systems, Standards, and Support

Now that you've seen the truth about caring, you can begin to rebuild with intention this quarter. Focus on how you deliver services—turning caring into everyday polices and processes, not just something reserved for exceptional moments.

THOUGHT-PROVOKING QUESTIONS

- Which systems do you have today that prevent your people from assisting customers?
- What areas of the organization have you accepted as inefficient and have therefore burnt out your employees?
- What do your incentives really incentivize?
- Where does technology help, and where does technology hinder?
- What skills are you not teaching that you should be teaching and developing?
- Do you teach your employees how to empathize, or just how to follow compliance?

THE 90-DAY CHALLENGE

- Rewrite the language of three of the most friction-filled policies and practices.
- Create an easy-to-use customer flow chart to accomplish something simple or change a complex process.
- Develop tools to assist agents, frontline employees, AI, and automation to help the customer and not destroy the experience.
- Train frontline employees AND leaders on empathy, judgement, and how caring can change the trajectory of the customer experience path.
- If needed, partner with BPO companies or other subject matter experts if needed and when necessary.
- Simplify one primary customer process.
- Align your performance metrics to care.

COMMITMENT TO CARING

- I will design and build systems for people, not against people.
- I will remove barriers rather than create excuses for bad behavior.
- I will invest in creating value, NOT JUST COMPLIANCE.
- I will design and create for DIGNITY and SIMPLICITY, not just efficiency.

QUARTER 3:
REIGNITE–BUILD A LASTING CULTURE

Focus: Leadership, Trust, and Belief

Culture is not what you say, but what you allow to happen. This quarter is about reigniting a sense of shared beliefs, both inside your organization and in the marketplace.

THOUGHT-PROVOKING QUESTIONS

- What behaviors do you demonstrate when someone experiences stress?
- Where have you chosen to remain silent rather than dealing with or addressing issues?
- Do people feel safe coming to you with bad news?
- Is there a trend toward greater trust in your workplace or are you losing that?
- Who feels like they are invisible in the workplace?
- Would you want your closest friends to work for this company/organization?

THE 90-DAY CHALLENGE

- Mentor five future or potential leaders.
- Recognize publicly across the organization employees who exhibit care each week.
- Directly address one toxic behavior.
- Restore psychological safety.
- Redesign onboarding based on purpose.
- Share stories of failure and lessons learned.
- Create a leadership shadowing opportunity.

COMMITMENT TO CARING

- I will lead openly and with vulnerability, openness, and trust.
- I will preserve trust at all costs.
- I will reward bravery over compliance.
- I will build a sense of community through action.

QUARTER 4:
REINVENT–LEAD TO LAST

Focus: Legacy, Innovation, and Future Readiness

This quarter is about ensuring that caring continues long after you are gone—building and creating organizations that are adaptable yet retain their heart and soul.

THOUGHT-PROVOKING QUESTIONS

- Are we planning for the next customer or the last customer?
- Will this culture exist if I'm no longer here?
- Are we investing in continuous learning for our employees and for our customers?
- Are we using AI and partnerships appropriately?
- What will we be remembered for?
- Are we building for five years or five quarters?

THE 90-DAY CHALLENGE

- Create a five-year CX roadmap.
- Create a plan to integrate AI and BPOs.
- Launch proactive pilot service projects.
- Develop succession plans.
- Create and institutionalize ongoing feedback mechanisms.
- Embed caring into organizational governance.
- Publish a company report on culture and trust.

COMMITMENT TO CARING

- I will build beyond myself.
- I will design for resilience, not reliance.
- I will innovate without losing our humanity.
- I will protect the flame.

Beckerman Caring Promise

Upon completing the Beckerman Caring Challenge, make the following commitment to yourself and to your followers:

- I will lead with humanity in a world that values efficiency above all else.
- I will treat customers as partners, not transactions.
- I will create systems that protect dignity.
- I will utilize technology to serve, not conceal.
- I will earn the trust of my customers and colleagues every day.
- I will leave this place better than I entered.

Caring is not a program; it is a lifestyle. Welcome to the Caring Movement.

"Listen and Learn from Everyone
but then Decide What Works Best for You."

—Natalie D. Beckerman

EPILOGUE:

MY PROMISE TO YOU

I didn't write this book so that you would be inspired for a day, a weekend, or until next week. I wrote it so that you could begin to lead differently—and think differently—for the rest of your life.

Over the past three decades, I've worked inside organizations: good, bad, and those in between. And I've watched the same thing happen time and again. Companies don't fail due to lack of strategy. Companies fail because they forget about people. Companies don't lose their way due to a lack of technology. Companies lose their way because they lose their connection to humanity. Companies don't fall behind because they can't compete. Companies fall behind because they stop caring.

I know this because I have been the person who made it happen. And it is the part of my career I am most sorry for.

For almost a year, my organization had been evaluating quality and insight automation platforms. Thirty-plus vendors. More than a hundred meetings. I had authored at least four formal proposal documents advocating for the shift. The market had matured. The platforms were sophisticated. Automation could review all interactions—not just a small sample—reduce scoring bias, increase the speed of coaching, and convert anecdotal insight into predictive intelligence. I believed it would make a material difference in our customer experience. I still believe that.

What I did not account for was what happened next.

One day, my boss simply stated: "Let's eliminate the entire quality assurance team. We're going to automate it anyway."

I was devastated. Not because I had doubts about the future—I didn't. But because I recognized the human impact of acting too quickly. There were over sixty quality assurance professionals who showed up every single day committed to doing their jobs with diligence, loyalty, and skill. They had built the framework we were now planning to modernize. I knew the magnitude of what this meant for them. I assured myself I was doing the responsible, forward-thinking thing for the organization.

I happened to be at one of our partner locations in South Africa when the final decision came down. I asked if I could delay delivering the news until I could do it in person when I returned home. HR and leadership were firm: It needed to happen immediately. I borrowed the CEO's office for a quiet and uninterrupted space. I logged onto the call.

Boom.

I advised the QA analysts that today would be their last day because we would be implementing quality automation.

I will never forget the silence. The shocked expressions. The confusion. And my own sorrow—because these were top-notch people. These were professionals who had shown up each and every day to defend our customers and our brand.

Then came the question I could not answer. The question that made me feel as if I had jumped out of an airplane with no way back:

"Why would you get rid of us now when the automation is not even installed?"

I supported the organizational position. "It is close. We are working on it." But in that moment, they were right. Why would we dismantle a functional system before the new system was available? Why not transition? Why not overlap? Why not maintain continuity instead of gambling with capability and credibility?

Following the end of that call, I sat in that office alone for nearly forty-five minutes with my face buried in my hands. I heard the rational voice: *We have saved nearly a million dollars*. Yet it didn't feel like a victory. It felt empty. Then the other voice came: *Maybe this was never going to happen. And I just destroyed sixty lives in less than thirty minutes.*

Nearly two years later, the organization failed to implement quality automation.

We eliminated a competent team based on a future that never existed.

The lesson was excruciating. Pursuing efficiency without executing discipline is not strategy—it is risk. Promising change prior to its implementation destroys trust at every level. And savings realized by eliminating a competent team in preparation for a future that never arrived creates vulnerability, not strength. Technological transformation needs to be implemented before employees are displaced. Confidence needs to be demonstrated before capability is removed.

This is why I wrote this book. Not to position myself as someone who always got it right—I didn't. But because I paid the price of not caring, in ways that still haunt me, and I refuse to let that cost go to waste.

I am tired. I am dissatisfied and frustrated that I thought I could create better customer experiences by building a new system—and sad that I created policies during my career that were meant to improve the customer experience but had the exact opposite effect. All of us make mistakes. The important thing is to learn from them.

This book is my promise to challenge those approaches and patterns—in your organization and in mine. The Beckerman Caring Challenge is not a theory, not a fad, and not a slogan. It is practical guidance designed to help you decide what is right. To help you show up for your people and your organization differently. To help you operate in a way that is better.

Because caring is not something you do once and move on from. Caring is something you recommit to each day—through meetings, through hiring, through technology, through recovery, through conflict, through growth, through crisis, through success.

At the beginning of this book, I asked you to carry five non-negotiables with you as you read. I ask you to carry them forward now:

Customers are partners, not transactions.
Employees are assets, not expenses.
Trust is capital, not collateral.
Culture is the foundation, not the façade.
Leadership is a service, not a title.

My promise to you is this: I will never ask you to choose between the performance of your organization and the well-being of your people. I will never tell you that caring is a trade-off for lower standards. Caring is raising the bar. Empathy is a superpower sharpened by discipline. And I will never pretend this work is easy. It isn't. But it is worth it.

If you elect to lead using the Beckerman Caring Challenge, you will be electing to create something greater than quarterly profits. You will be creating trust. You will be creating faith. You will be creating a legacy.

And I promise to stand with you in that work.

Natalie D. Beckerman

ENDNOTES

1 Osorio-Mazilli, Gabriel. "As Peloton Shares Fall over 90%, a Comeback Strategy Surges." *Entrepreneur*, June 13, 2023. https://www.entrepreneur.com/finance/as-peloton-shares-fall-over-90-a-comeback-strategy-surges/454064.

2 NPS Prism (Bain & Company and J.D. Power), reported in "USAA, Navy Federal Lead Consumer Banking NPS Rankings," CX *Dive*, February 23, 2024, https://www.customerexperiencedive.com/news/usaa-navy-federal-consumer-banking-nps-rankings/708295/.

3 Reichheld, Frederick F., and W. Earl Sasser Jr. "Zero Defections: Quality Comes to Services." *Harvard Business Review* 68, no. 5 (September–October 1990): 105–111. For updated analysis, see Reichheld, Frederick F., and Rob Markey. The Ultimate Question 2.0: How Net Promoter Companies Thrive in a Customer-Driven World. Boston: Harvard Business Review Press, 2011.

4 NPS Prism, CX *Dive*.

5 The quote "People will forget what you said, people will forget what you did, but people will never forget how you made them feel" is widely attributed to Maya Angelou; however, no confirmed written source in her published work has been identified. The attribution is widely accepted and commonly cited in leadership and customer experience literature.

6 Consumer Intelligence Research Partners (CIRP), "Amazon Prime Member Spending Is Even More Important Than We Thought," *CIRP Amazon Report*, January 28, 2025, https://cirpamazon.substack.com/p/amazon-prime-member-spending-is-even. As of 2024: Prime members spend approximately $1,170/year vs. $570 for non-Prime members.

7 Gartner, "Rethink Your Customer Service Strategy to Drive Self-Service," *Smarter With Gartner*, https://www.gartner.com/smarterwithgartner/rethink-customer-service-strategy-drive-self-service (70% of customers use self-service at some point in their resolution journey). See also Forrester Research: 67% of consumers prefer self-service over speaking to a representative for simple inquiries.

ACKNOWLEDGMENTS

This book is the culmination of a lifetime of learning, loving, failing, and growing, of being challenged, supported, and believed in, often when I needed it most.

I want to express my sincere appreciation to Anouk, for your love of the English language, for your ability to teach me the distinction between proper English and American English, and for your commitment to pronunciation, grammar, spelling, and writing. But more importantly, thank you for believing in me, encouraging me, and serving as my largest supporter, fan, and cheerleader. Your belief in me has inspired mine, and that's what carries over into everything I do.

Thank you to my parents for your lifelong love, care, and unconditional support. You instilled in me the conviction that I could accomplish anything, but more importantly, the courage to attempt it. The foundation you've given me has impacted each chapter of my life and every page of this book. Dad, you would have loved this book.

Thank you to my friends, colleagues, family, and teammates for standing by my side, lifting me up, and helping me remember who I am when even for me, it got too much. Your encouragement and concern, particularly during the writing of this book, provided me with the strength to continue. I'll always appreciate your loyalty, your honesty, and your trust in my purpose.

Thank you to my publisher, Davin Salvagno, my literary agent, Scott Miller, and Maison Vero for your partnership, guidance, and trust. This process has taken me far beyond where I expected. Through it, I've learned a lot more about my strengths, my areas of improvement, and my tenacity than I have in quite some time. Your support and endless encouragement enabled me to take my vision and make it a reality. For that, I will always be grateful.

To all the various bosses and leaders (good and bad) whose careers have formed mine, thank you. Each of your experiences has taught me something. For me, there is no end to learning. There is no end to change. There is no end to growth. And regardless of whether you knew it or not, each of you has forced me to push myself to grow a little more every day and to continue as my relentless and tenacious self. I am stronger as a result.

To everyone who has worked with me, for me, and alongside me on my teams: You have taught me more than you will ever realize. You have shown me that leadership isn't just about results, numbers, and outcomes. It's about people, about listening, about leading, about mentoring, about coaching, and about caring. You have helped me understand that true success comes from building relationships based on trust and humanity, not fear and control.

Finally, to you, the reader: Thank you for selecting this book. Thank you for opening yourself up to this message. Thank you for adopting the Beckerman Caring Challenge. My wish is that these pages will motivate you to lead differently, to care more profoundly, and to act more bravely.

Together, let's ignite the flame again. Let's bring caring back into the workplace in a way that has never been experienced or felt by anyone.

ABOUT THE AUTHOR

Natalie Beckerman is the world's top authority on customer experience and service transformation, frequently known as the Global Customer Experience Connoisseur.

She combines efficient business processes with human-centered design to show that human-centered leadership is not just a nice-to-have soft asset, but a key differentiator for competitiveness. A catalyst of change and a visionary / executioner, she has redesigned complex ecosystems, restored damaged cultures, and changed how companies operate. She merges high-performance operations with behavioral science and operational discipline to ensure that strategy is implemented and measurable.

As Executive Vice President and Chief Business Officer at iQor, a leading provider of global customer experience business process outsourcing, Natalie leads iQor's international growth and transformation. In addition to this role, Natalie founded COG LLC (Customer Operations Group), a boutique consultancy focused on providing courageous, operationally-driven, and customer-centric leadership.

Prior to iQor, Natalie served as global head of Customer Support Operations at IHG Hotels & Resorts, has held senior roles at Rocket Mortgage and Leaf Home, and was the EVP/COO at Fidelity National Home Warranty. Her experience spans hospitality, fintech, telecom, and home services, both domestic and internationally.

Natalie is a highly sought-after speaker, advisor, and coach. As such, she sits on the Customer Contact Week Advisory Board, is a CCWomen Hall of Fame inductee, and one of CX Networks top 30 contact center leaders in 2025.

Natalie holds an MBA from the University of Bath, an MS from Michigan State University, and a BA from Northwestern University.

When Did You Stop Caring?—her life's work—calls leaders to light the flame of humanity before efficiency extinguishes it.

BOOK NATALIE BECKERMAN

TO SPEAK WITH YOUR TEAM, PROGRAM, OR ORGANIZATION

For speaking engagements, Natalie is represented by the Gray & Miller Agency, a leading speaking firm comprised of the world's most diverse, engaging, and influential speakers, authors, and thought leaders.

For speaking inquiries please visit **GrayMillerAgency.com**.

For consulting engagements, reach out to **Natalie@NatalieBeckerman.com**.

To buy more copies of this book go to **NatalieBeckerman.com**.

To request information on how to get in touch with Natalie, reach out to **Natalie@NatalieBeckerman.com**.

MAISON VERO

Representing a community of authors whose books have collectively sold hundreds of millions of copies, the founders of The Gray + Miller Agency launched Maison Vero, a professional publishing house that partners with rising authors to bring their thought leadership to the world. Our representation covers every aspect of thought leadership, including U.S. senators, governors, and ambassadors, billionaire founders and entrepreneurs, researchers, academics, scientists, consultants, practitioners, social influencers, C-suite leaders, adventurers, professional athletes, artists, and creators. We partner with thought leaders and world changers like you who have a story to tell. By bringing decades of professional expertise to our clients, we are charting a new path in a timeless industry that transcends publishing norms, transforming powerful thoughts into impactful books that inspire minds, ignite hearts, and open doors.

Visit maisonvero.com to view our growing list of authors, or to submit a proposal for publication consideration.

Follow Maison Vero for insight and inspiration on social media:

in MaisonVero f MaisonVero MaisonVeroPublishing

For information about special discounts for bulk purchases, please call 1-949-333-4872 or email info@graymilleragency.com.

Maison Vero is a partner brand of The Gray + Miller Agency, a speaking, literary, and talent consortium. For more information on the talent represented by The Gray + Miller Agency, or to bring any of our thought leaders to your organization or live event, please visit our website at **graymilleragency.com**.

www.ingramcontent.com/pod-product-compliance
Lightning Source LLC
LaVergne TN
LVHW091514200926
848483LV00008B/461/J